Springer INdAM Series

Volume 36

Editor-in-Chief

Giorgio Patrizio, Università di Firenze, Florence, Italy

Series Editors

Claudio Canuto, Politecnico di Torino, Turin, Italy
Giulianella Coletti, Università di Perugia, Perugia, Italy
Graziano Gentili, Università di Firenze, Florence, Italy
Andrea Malchiodi, SISSA - Scuola Internazionale Superiore di Studi Avanzati, Trieste, Italy
Paolo Marcellini, Università di Firenze, Florence, Italy
Emilia Mezzetti, Università di Trieste, Trieste, Italy
Gioconda Moscariello, Università di Napoli "Federico II", Naples, Italy
Tommaso Ruggeri, Università di Bologna, Bologna, Italy

Springer INdAM Series

This series will publish textbooks, multi-authors books, thesis and monographs in English language resulting from workshops, conferences, courses, schools, seminars, doctoral thesis, and research activities carried out at INDAM - Istituto Nazionale di Alta Matematica, http://www.altamatematica.it/en. The books in the series will discuss recent results and analyze new trends in mathematics and its applications.

THE SERIES IS INDEXED IN SCOPUS

More information about this series at http://www.springer.com/series/10283

Marco Donatelli · Stefano Serra-Capizzano
Editors

Computational Methods for Inverse Problems in Imaging

 Springer

Editors
Marco Donatelli
Department of Science and High
Technology
University of Insubria
Como, Italy

Stefano Serra-Capizzano
Department of Humanities and Innovation
University of Insubria
Como, Italy

ISSN 2281-518X   ISSN 2281-5198   (electronic)
Springer INdAM Series
ISBN 978-3-030-32884-9   ISBN 978-3-030-32882-5   (eBook)
https://doi.org/10.1007/978-3-030-32882-5

This Springer imprint is published by the registered company Springer Nature Switzerland AG
The registered company address is: Gewerbestrasse 11, 6330 Cham, Switzerland

Preface

In recent years, mathematical research in the field of inverse problems for imaging has attracted great interest, receiving a significant boost and producing significant advances for real-life applications. Indeed, despite the huge improvement in the hardware imaging systems, analysis of the imaging models combined with study of appropriate numerical methods and their efficient implementation is of fundamental importance for innovative imaging applications arising from very different domains of applied science, such as medical imaging, microscopy, astronomy, and seismology.

As is well known, several image restoration problems are ill-posed inverse problems. As a consequence, discretization of the equations relating the unknown solution to the data leads to problems affected by numerical instability. Therefore, the formulation of such inverse problems requires accurate mathematical modeling, including a statistical model of the noise affecting the data, and additional constraints on the solution. Furthermore, the introduction of regularization techniques in the model enables one to guide the reconstruction method toward a solution with meaningful features. As a result, the solution of inverse problems is in general reduced to the constrained minimization of suitable functionals with special structure. The computational methods to efficiently address these problems are based on new optimization algorithms, on the use of (numerical) structured linear algebra, and on fast multilevel techniques, such as wavelets and framelets.

In recent years, optimization-based tools for smooth and nonsmooth convex minimization problems have been used to solve challenging imaging problems. These tools are based on first-order methods, also with variable metrics, and related acceleration techniques, dual or primal-dual approaches, and Bregman-type schemes. The need to satisfy certain modeling aspects, such as the robustness to noise and the recovery of sparse and discontinuous signals, requires the development of numerical methods for nonconvex problems. Moreover, the huge amount of data requires fast convergent methods with a low computational cost, for instance, working in subspaces of small dimensions or introducing acceleration techniques based on regularizing preconditioners.

Some applications requiring recovery of special features in the images have, during the last decade, given rise to schemes that rely on the notion of sparsity, that is, the idea that data can be approximated using a relatively small number of functions from an appropriate representation basis. By taking advantage of multilevel techniques based on wavelet or shearlet decomposition, robust algorithms can be devised for reconstruction starting from incomplete and/or truncated data. Several current and challenging applications are to be found in astronomy (e.g., blind deconvolution of interferometric images of the Large Binocular Telescope Interferometer), microscopy (e.g., reconstruction of multiple images in STED microscopy or images acquired by means of differential interference contrast microscopy), and biomedical imaging [e.g., reconstruction of a small region of interest from truncated computed tomography (CT) projection data, reconstruction of images from microwave systems, and reconstruction methods in PET and MRI imaging in functional medicine].

This book includes both contributions on numerical methods based on the most recent optimization tools and contributions on challenging inverse problems in imaging. In particular, there is a review of variable metric first-order methods and regularizing preconditioners for image deblurring combined with the multilevel framelet decomposition. The structure of the preconditioner, like Toeplitz matrices, is crucial, and the spectral distribution of Toeplitz-function sequences is also analyzed. The considered applications include multiple image deblurring, segmentation of astronomical images, and CT for bone tissue.

Como, Italy   Marco Donatelli
September 2019   Stefano Serra-Capizzano

Contents

Recent Advances in Variable Metric First-Order Methods 1
Silvia Bonettini, Federica Porta, Marco Prato, Simone Rebegoldi,
Valeria Ruggiero and Luca Zanni

Structure Preserving Preconditioning for Frame-Based Image Deblurring 33
Davide Bianchi, Alessandro Buccini and Marco Donatelli

Non-stationary Structure-Preserving Preconditioning for Image Restoration 51
Pietro Dell'Acqua, Marco Donatelli and Lothar Reichel

Numerical Investigation of the Spectral Distribution of Toeplitz-Function Sequences 77
Sean Hon and Andy Wathen

The Hough Transform and the Impact of Chronic Leukemia on the Compact Bone Tissue from CT-Images Analysis 93
Anna Maria Massone, Cristina Campi, Francesco Fiz
and Mauro Carlo Beltrametti

Multiple Image Deblurring with High Dynamic-Range Poisson Data 117
Marco Prato, Andrea La Camera, Carmelo Arcidiacono, Patrizia Boccacci
and Mario Bertero

On the Segmentation of Astronomical Images via Level-Set Methods 141
Silvia Tozza and Maurizio Falcone

About the Editors

Marco Donatelli is an Associate Professor of Numerical Analysis at the Department of Science and High Technology, University of Insubria (Italy). He was awarded a Ph.D. in Applied Mathematics by the University of Milan in 2006. His research interests include regularization methods of inverse problems, preconditioning and multigrid methods for structured matrices. He is author of more than 70 papers and he serves on the editorial boards of three international journals.

Stefano Serra-Capizzano is a Full Professor of Numerical Analysis at the Department of Humanities and Innovation, Deputy Rector of the University of Insubria (Italy), and long-term Visiting Scholar at Uppsala University (Sweden). He has authored over 200 research papers in different areas of mathematics, including numerical linear algebra, spectral theory, approximation theory, and inverse problems, with more than 100 collaborators around the globe. He is the founder of the Ph.D. Program "Mathematics of Computation" and of the Department of Science and High Technology at the University of Insubria.

Recent Advances in Variable Metric First-Order Methods

Silvia Bonettini, Federica Porta, Marco Prato, Simone Rebegoldi, Valeria Ruggiero and Luca Zanni

Abstract Minimization problems often occur in modeling phenomena dealing with real-life applications that nowadays handle large-scale data and require real-time solutions. For these reasons, among all possible iterative schemes, first-order algorithms represent a powerful tool in solving such optimization problems since they admit a relatively simple implementation and avoid onerous computations during the iterations. On the other hand, a well known drawback of these methods is a possible poor convergence rate, especially showed when an high accurate solution is required. Consequently, the acceleration of first-order approaches is a very discussed field which has experienced several efforts from many researchers in the last decades. The possibility of considering a variable underlying metric changing at each iteration and aimed to catch local properties of the starting problem has been proved to be effective in speeding up first-order methods. In this work we deeply analyze a possible way to include a variable metric in first-order methods for the minimization of a functional which can be expressed as the sum of a differentiable term and a non-

Members of the INdAM research group GNCS.

S. Bonettini (✉) · F. Porta · M. Prato · L. Zanni
Dipartimento di Scienze Fisiche, Informatiche e Matematiche,
Università degli Studi di Modena e Reggio Emilia, Via Campi 213/b, 41125 Modena, Italy
e-mail: silvia.bonettini@unimore.it

F. Porta
e-mail: federica.porta@unimore.it

M. Prato
e-mail: marco.prato@unimore.it

L. Zanni
e-mail: luca.zanni@unimore.it

S. Rebegoldi · V. Ruggiero
Dipartimento di Matematica e Informatica, Università degli Studi di Ferrara,
Via Saragat 1, 44121 Ferrara, Italy
e-mail: simone.rebegoldi@unife.it

V. Ruggiero
e-mail: valeria.ruggiero@unife.it

© Springer Nature Switzerland AG 2019 1
M. Donatelli and S. Serra-Capizzano (eds.), *Computational Methods for Inverse Problems in Imaging*, Springer INdAM Series 36,
https://doi.org/10.1007/978-3-030-32882-5_1

differentiable one. Particularly, the strategy discussed can be realized by means of a suitable sequence of symmetric and positive definite matrices belonging to a compact set, together with an Armijo-like linesearch procedure to select the steplength along the descent direction ensuring a sufficient decrease of the objective function.

Keywords Constrained optimization · Gradient projection methods · Convex optimization

1 Introduction

The class of problems investigated in this work can be written as

$$\min_{x \in \mathbb{R}^n} f(x) \equiv f_0(x) + f_1(x) \tag{1}$$

where f_0 is continuously differentiable on an open subset Ω_0 of $\mathbb{R}^n$ containing $\mathrm{dom}(f_1) = \{x \in \mathbb{R}^n : f_1(x) < +\infty\}$ and f_1 is a proper, convex, lower semicontinuous and bounded from below function. Moreover we suppose that the set $\mathrm{dom}(f_1)$ is closed.

Optimization model (1) is considerably relevant, since it is related to several problems arising from real-life applications like image and signal processing, compressed sensing, machine learning (see for example [8, 9, 41, 62, 71]). However, such problems often concern with large or high-dimensional datasets and their solutions need to be processed quickly. First-order methods are effective tools for facing minimization problems of this kind, thanks to their simplicity of implementation, limited storage requirements and low computational cost per iteration. In the family of first-order algorithms, forward-backward (FB) schemes [26] are especially tailored for the nature of the objective function in (1), since they consist of a forward step, which exploits the differentiability of f_0, and a backward step, which takes advantage of the convexity of f_1. More in detail, the general FB iteration can be expressed as

$$\begin{aligned} z^{(k)} &= x^{(k)} - \alpha_k \nabla f_0(x^{(k)}) \\ y^{(k)} &= \mathrm{prox}_{\alpha_k f_1}(z^{(k)}) \\ d^{(k)} &= y^{(k)} - x^{(k)} \\ x^{(k+1)} &= x^{(k)} + \lambda_k d^{(k)} \end{aligned} \tag{2}$$

where λ_k and α_k are suitable positive parameters and the proximal operator $\mathrm{prox}_h(\cdot)$ associated to a general convex function h is defined by

$$\mathrm{prox}_h(x) = \operatorname*{argmin}_{w \in \mathbb{R}^n} \frac{1}{2} \|w - x\|^2 + h(w) \,. \tag{3}$$

We recall that, if the function f_1 is the indicator function of a closed and convex set C, i.e.,

$$f_1(x) = \iota_C(x) = \begin{cases} 0 & x \in C \\ +\infty & x \notin C \end{cases}$$

then algorithm (2) reduces to the class of standard gradient projection methods [10, Chap. 2] since, in this case, the proximal operator associated to f_1 consists of a projection onto the set C.

Despite of their merits, classical FB algorithms can exhibit an unsatisfactory behaviour in terms of convergence rate, especially when a particularly accurate solution is required. For this reason, the recent literature suggested some techniques to accelerate FB methods by exploiting two principal ingredients: an underlying variable metric different from the usual Euclidean one and/or an extrapolation step which employs the information from the two previous iterations.

In the framework of differentiable constrained optimization problems, diagonally scaled gradient directions have been often successfully exploited in order to accelerate the performance of classical gradient projection methods [31, 48, 49, 52, 64, 72, 77]. The possibility of scaling the gradient directions also in FB algorithms led to the class of variable metric FB methods. The general scheme of a variable metric FB approach has been studied by several authors [24, 27, 28, 38, 69] in the following form

$$z^{(k)} = x^{(k)} - \alpha_k D_k^{-1} \nabla f_0(x^{(k)})$$
$$y^{(k)} = \mathrm{prox}_{\alpha_k f_1}^{D_k}(z^{(k)})$$
$$d^{(k)} = y^{(k)} - x^{(k)} \tag{4}$$
$$x^{(k+1)} = x^{(k)} + \lambda_k d^{(k)}$$

where $\{D_k\}_{k \in \mathbb{N}} \subset \mathbb{R}^{n \times n}$ is a sequence of symmetric and positive definite scaling matrices whose aim is capturing some local features of the minimization problem (1) without introducing significant additional computational costs. In this context, the proximity operator is generalized as

$$\mathrm{prox}_{\alpha_k f_1}^{D_k}(x) = \underset{w \in \mathbb{R}^n}{\mathrm{argmin}} \, \frac{1}{2\alpha_k}(w - x)^T D_k(w - x) + f_1(w). \tag{5}$$

It is worth to notice that, even if the variable metric FB method considered in the above mentioned papers is formally described by iteration (4), different choices of the parameters λ_k, α_k and D_k lead to substantially different algorithms in terms of convergence properties and practical implementation. Indeed, although the theoretical convergence rate on the objective function values for variable metric FB algorithms is at most sublinear (i.e. $f(x^{(k)}) - f^* = \mathcal{O}(1/k)$, where f^* is the optimal value of the objective function) [7, 14, 38, 69], it has been numerically showed [14, 60] that a suitable combination of the steplength parameter α_k and the scaling operator D_k can allow method (4) to reach much better practical performances (comparable with those of schemes with proved superlinear convergence rate). In Sects. 3 and 4 we discuss a clever way to select parameters λ_k, α_k, D_k in the framework of FB algorithms. In particular, our approach

- allows a certain freedom of choice for α_k and D_k which have simply to belong to compact sets;
- guarantees the theoretical convergence properties thanks to the computation of the parameter λ_k by means of an Armijo-like backtracking procedure.

This selecting recipe is one of the main features which characterize our approach (and, in our opinion, also one of its main strength): for instance, in [24] the steplength α_k is constant during the iterations and the scaling matrix D_k has to be selected by following a Majorization—Minimization approach; in [28, 38] the computation of α_k and/or λ_k and the spectrum of D_k is related to the Lipschitz constant of ∇f_0 (which, unfortunately, is not always known).

Another important issue faced in the literature on FB methods is the computation of the proximal point [3, 24, 68, 75]: indeed the solution of the minimization problem associated to the definition of the proximal operator (5) is not always known in a closed form. In Sect. 4.3 we detail a strategy to compute an approximation of the proximal point and its practical implementation which preserves all the theoretical convergence properties.

As we have already mentioned, another technique to accelerate FB algorithms can be realized by adding an extrapolation step (also called inertial force) to the standard FB scheme (2). Originally suggested by Nesterov [55] for gradient methods applied to differentiable optimization problems, this idea has been borrowed by Beck and Teboulle in [7], where they propose a FB method with extrapolation in the following form

$$
\begin{aligned}
w^{(k)} &= x^{(k)} + \beta_k(x^{(k)} - x^{(k-1)}) \\
z^{(k)} &= w^{(k)} - \alpha_k \nabla f_0(w^{(k)}) \\
x^{(k+1)} &= \mathrm{prox}_{\alpha_k f_1}(z^{(k)})
\end{aligned}
\tag{6}
$$

also known in the literature as Fast Iterative Shrinkage-Thresolding Algorithm (FISTA). The positive steplength parameters α_k and β_k are conveniently chosen in order to ensure the convergence of the algorithm. It is worth to observe that, since β_k in (6) is positive, $w^{(k)}$ is not a convex combination of $x^{(k)}$ and $x^{(k-1)}$, but an extrapolation of these iterates. The convergence properties of FISTA have been extensively analyzed in the last ten years. More in detail, under the assumption that both f_0 and f_1 are convex, the following results hold for FISTA:

In [7] the authors show a subquadratic convergence rate on the objective function values, namely $f(x^{(k)}) - f^* = \mathcal{O}(1/k^2)$, where f^* is the optimal value of the objective function;

In [70] the quadratic convergence rate is also guaranteed when an inexact computation of the proximal point is considered;

In [23] Chambolle and Dossal prove the convergence of the sequence $\{x^{(k)}\}_{k \in \mathbb{N}}$;

In [4]an improved $o(1/k^2)$ convergence rate result is shown together with a simpler proof for the convergence of the iterates.

A slightly different version of FB method with extrapolation has been suggested in [57] where the authors generalize the so called heavy-ball method proposed by Polyak [58] and prove the convergence of their generalized method for nonconvex objective functions. The corresponding update scheme is

$$
\begin{aligned}
w^{(k)} &= \beta_k (x^{(k)} - x^{(k-1)}) \\
z^{(k)} &= x^{(k)} - \alpha_k \nabla f_0(x^{(k)}) + w^{(k)} \\
x^{(k+1)} &= \text{prox}_{\alpha_k f_1}(z^{(k)}).
\end{aligned}
\tag{7}
$$

We remark that the generalized heavy-ball method (7) uses gradients based on the current iterate, while the inertial FB algorithm based on the Nesterov's approach (6) evaluates the gradient at the extrapolated points. As for the original versions, Nesterov's accelerated gradient method is faster than the heavy-ball one on weakly convex functions [37]. In Sect. 5 we discuss the possibility of considering a variable metric FB method with extrapolation: starting from scheme (6) we investigate how to combine the inertial idea with a variable metric. If the functional to minimize is convex, the convergence of the resulting method is ensured also in case of inexact computation of the proximal point.

The FB scheme can be extended when f_0 is convex but nondifferentiable. In this case, the gradient of f_0 in (2) can be replaced by a subgradient or by an approximation of it. The corresponding method is known in the literature as *proximal subgradient method* [30, 50, 73]. Even if the proximal subgradient method can be formally framed in the iteration (2), there are substantial differences with respect to the differentiable case, especially in the selection strategies for the stepsize parameters. On the other side, variable metric techniques can be introduced also in this context, leading to a variable metric proximal subgradient method which is discussed in Sect. 6.

To summarize, the aim of this paper is to investigate how to accelerate FB methods (with and without extrapolation) by accounting a variable metric instead of the usual fixed Euclidean one. A possibility to achieve this goal consists of considering a suitable sequence of symmetric and positive definite matrices which modify both the forward step (by scaling the descent directions) and the backward step (by computing the proximal point in the norm induced by the scaling matrix itself).

2 Split Gradient Methods: The Starting Point

The starting point of the ideas that will be presented in the following sections lies in two inspiring papers by Lantéri et al. [48, 49]. In these works, by considering the first-order optimality Karush–Kuhn–Tucker (KKT) conditions, the authors devised a scaled gradient method for the solution of a simplified version of problem (1), namely

$$\min_{x \in \mathbb{R}^n} f_0(x) + \iota_{x \geq 0}(x) \tag{8}$$

where f_0 is a continuously differentiable, coercive, convex function and $\iota_{x \geq 0}$ is the indicator function of the non-negative orthant. Since the objective function in (8) is convex, all its minima are global and, in particular, a point x^* is a solution if and only if the KKT conditions are verified at x^*:

$$x^* \cdot \nabla f_0(x^*) = 0 \tag{9}$$
$$x^* \geq 0, \ \nabla f_0(x^*) \geq 0, \tag{10}$$

where $\cdot$ denotes the component-wise product.

The key point to obtain the desired algorithm is to consider a proper decomposition of the gradient of f_0 in a non-negative part and a positive one:

$$-\nabla f_0(x) = U(x) - V(x) \quad \text{where} \quad U(x) \geq 0, \ V(x) > 0. \tag{11}$$

We remark that, even if such a decomposition is not unique, it always exists. Thanks to (11), the first KKT condition (9) can be rewritten as a fixed point equation

$$x^* = A(x^*)$$

where

$$A(x) = x \cdot \frac{U(x)}{V(x)}$$

and the fraction symbol indicates component-wise division. The operator A

- is well defined since $V(x) > 0$, for any feasible x,
- is continuous since f_0 is continuously differentiable,
- is not a contraction in general.

Given $x^{(0)} > 0$ and by applying the method of successive approximations, the following algorithm can be considered

$$x^{(k+1)} = x^{(k)} \cdot \frac{U(x^{(k)})}{V(x^{(k)})}. \tag{12}$$

It is possible to include scheme (12) in the class of scaled gradient method: indeed by summing and subtracting $x^{(k)}$ to the second member of (12) we obtain the following equalities:

$$x^{(k+1)} = x^{(k)} + x^{(k)} \cdot \left(\frac{U(x^{(k)})}{V(x^{(k)})} - 1 \right),$$

$$x^{(k+1)} = x^{(k)} + x^{(k)} \cdot \left(\frac{U(x^{(k)}) - V(x^{(k)})}{V(x^{(k)})} \right),$$

$$x^{(k+1)} = x^{(k)} - x^{(k)} \cdot \left(\frac{\nabla f_0(x^{(k)})}{V(x^{(k)})} \right),$$

$$x^{(k+1)} = x^{(k)} - D_k^{-1} \nabla f_0(x^{(k)}),$$

where

$$D_k^{-1} = \operatorname{diag} \left(\frac{x^{(k)}}{V(x^{(k)})} \right). \tag{13}$$

We remark that all the iterates $x^{(k)}$ generated by this approach automatically satisfy the non-negative constraints. However, since A is not a contraction, the convergence can not be ensured for the algorithm written in this form. For this reason, in [48, 49] the authors add a steplength multiplying the scaled gradient direction in order to obtain the wished convergence. Particularly, the revised algorithm can be presented as

$$x^{(k+1)} = x^{(k)} - \alpha_k \frac{x^{(k)}}{V(x^{(k)})} \left(V(x^{(k)}) - U(x^{(k)}) \right), \tag{14}$$

where α_k can be computed by means of a suitable backtracking procedure in the interval $\left(0, \alpha_k^{(0)} \right]$ with $\alpha_k^{(0)}$ conveniently selected to ensure $x^{(k+1)} \geq 0$.

Due to the gradient decomposition idea (11) at the basis of the algorithm just described, the resulting scheme is known in the literature as split gradient method (SGM). The SGM has been developed by Lantéri et al. as a generalization of two iterative approaches employed to solve minimization problems in the field of image processing. The next subsection is devoted to explain all the details.

2.1 Classical Reconstruction Algorithms in Imaging

Variational approaches to image restoration [76] suggest to recover the unknown object through iterative schemes suited for the constrained minimization problem (8) where, in this case, f_0 measures the discrepancy of a given image $x \in \mathbb{R}^n$ from the observed data $y \in \mathbb{R}^n$. The definition of the function f_0 depends on the noise type introduced by the acquisition system. Particularly, in the case of additive white Gaussian noise the cost function is characterized by a least squares distance of the form

$$f_0(x) = f_0^{LS}(x) = \frac{1}{2} \|Hx + bg - y\|^2 \tag{15}$$

where $H \in \mathbb{R}^{n \times n}$ is a typically ill-conditioned matrix describing a blurring effect and bg is a known non-negative background radiation. The matrix H can be considered with non-negative entries, generally dense and such that $\sum_{i=1}^{n} H_{ij} > 0 \; \forall j$ and $\sum_{j=1}^{n} H_{ij} > 0 \; \forall i$. For (15), it holds that

$$- \nabla f_0^{LS}(x) = U^{LS}(x) - V^{LS}(x) = H^T y - (H^T H x + bg). \tag{16}$$

On the other hand, when the noise affecting the data is of Poisson type, the so-called Kullback–Leibler (KL) divergence is used:

$$f_0^{KL}(x) = \sum_{i=1}^{n} \left\{ y_i \ln \frac{y_i}{(Hx + bg)_i} + (Hx + bg)_i - y_i \right\} \tag{17}$$

where we assume that $0 \ln 0 = 0$ and $(Hx + bg)_i > 0$, for any i such that $y_i \neq 0$. In this second case the gradient of f_0^{KL} enjoys the following decomposition

$$- \nabla f_0^{KL}(x) = U^{KL}(x) - V^{KL}(x) = H^T \frac{y}{Hx + bg} - H^T 1 \tag{18}$$

where $1 \in \mathbb{R}^n$ is a vector whose components are all equal to one.

The iterative space reconstruction algorithm (ISRA) has been introduced in [31] to face problem (8) in the case of Gaussian noise. The sequence $\{x^{(k)}\}_{k \in \mathbb{N}}$ generated by ISRA is defined by

$$x^{(k+1)} = x^{(k)} \frac{H^T y}{H^T H x^{(k)} + bg}. \tag{19}$$

It is simple to verify that this last iteration is equivalent to

$$x^{(k+1)} = x^{(k)} - D_k^{-1}(H^T H x^{(k)} + bg - H^T y), \quad D_k^{-1} = \text{diag}\left(\frac{x^{(k)}}{H^T H x^{(k)} + bg} \right)$$

which is a particular case of algorithm (14) where α_k is always equal to 1. The asymptotic convergence of (19) has been proved in [35] when $bg = 0$, but the proof of convergence can be easily extended to the case $bg \neq 0$.

As for the Poisson noise, a classical approach to minimize (17) under non-negative constraints is the expectation maximization (EM) algorithm proposed in [72] and known as Richardson-Lucy (RL) algorithm in image deconvolution [52, 64]. The EM scheme generates a sequence of iterates in the following way

$$x^{(k+1)} = \frac{x^{(k)}}{H^T 1} H^T \frac{y}{Hx^{(k)} + bg} \tag{20}$$

which can be equivalently rewritten as

$$x^{(k+1)} = x^{(k)} - D_k^{-1} \left(H^T 1 - H^T \frac{y}{Hx^{(k)} + bg} \right), \quad D_k^{-1} = \text{diag} \left(\frac{x^{(k)}}{H^T 1} \right).$$

It is possible to conclude that also the EM algorithm belongs to the class of split gradient methods (14) with $\alpha_k = 1$. In the case of $bg = 0$ several convergence proofs of the algorithm (20) are available [46, 47, 53]. Very recently [67], the convergence of the EM approach has been provided also for $bg \neq 0$.

To conclude this section we remark that the papers by Lanteri et al. suggest a practical way to define scaled gradient methods which exploits the nature of the problem to minimize. Although the proposed approach is very promising, it appears limited since it only considers non-negative constraints and differentiable and convex objective functions. In the next sections we employ the gradient decomposition idea in first-order algorithms able to solve more general optimization problems.

3 A Scaled Gradient Projection Method

In order to solve minimization problem of kind (8) but equipped with more general constraints, i.e.,

$$\min_{x \in \mathbb{R}^n} f_0(x) + \iota_\Omega, \quad \Omega \subset \mathbb{R}^n \text{ closed, convex,} \tag{21}$$

a scaled gradient projection (SGP) method has been proposed in [12]. This SGP algorithm can be considered a generalization of the split gradient method (14) and, at the same time, a particular approach belonging to the class of variable metric FB methods (4). The basic SGP scheme is given by

$$\begin{aligned}
z^{(k)} &= x^{(k)} - \alpha_k D_k^{-1} \nabla f_0(x^{(k)}) \\
y^{(k)} &= \mathbb{P}_\Omega^{D_k}(z^{(k)}) \\
d^{(k)} &= y^{(k)} - x^{(k)} \\
x^{(k+1)} &= x^{(k)} + \lambda_k d^{(k)}
\end{aligned} \tag{22}$$

where

- α_k is the steplength parameter chosen in a fixed range $[\alpha_{\min}, \alpha_{\max}]$ with $0 < \alpha_{\min} < \alpha_{\max}$;
- D_k is a symmetric and positive definite scaling matrix with all the eigenvalues contained in $[\frac{1}{\mu}, \mu]$ with $\mu \geq 1$; hereafter the set of matrices satisfying these properties will be indicated by $\mathcal{D}_\mu$. If $D_k \in \mathcal{D}_\mu$, then D_k^{-1} also belongs to $\mathcal{D}_\mu$;
- λ_k is computed by means of a backtracking procedure over the interval $(0, 1]$ to ensure a sufficient decrease of the objective function. A classical example of linesearch is the well-known Armijo rule [10, Sect. 2.3]: for given scalars

$0 < \beta, \sigma < 1$, the parameter λ_k is set equal to β^{m_k}, where m_k is the first non-negative integer m for which

$$f_0(x^{(k)}) - f_0(x^{(k)} + \beta^m d^{(k)}) \geq -\sigma \beta^m \nabla f_0(x^{(k)})^T d^{(k)}. \tag{23}$$

A non-monotone version of (23) has been suggested in [43]: instead of $f_0(x^{(k)})$ in the first member of the previous inequality, it is possible to consider the maximum value of the objective function in the last M iterations, namely $f_{\max} = \max_{0 \leq j \leq \min(k, M-1)} f(x^{(k-j)})$;

$\mathbb{P}_{\Omega}^{D_k}(\cdot)$ represents the projection operator onto Ω with respect to the norm induced by D_k:

$$\mathbb{P}_{\Omega}^{D_k}(z^{(k)}) = \underset{x \in \Omega}{\arg\min} \, \nabla f_0(x^{(k)})^T(x - x^{(k)}) + \frac{1}{2\alpha_k}(x - x^{(k)})^T D_k(x - x^{(k)}).$$

We underline that the vector $d^{(k)}$ results a descent direction for the function f_0 at $x^{(k)}$, that is, $\nabla f_0(x^{(k)})^T d^{(k)} < 0$, unless $x^{(k)}$ is a stationary point for (21).

Typically, the SGP method is applied when the projection onto Ω is computable by either a closed formula or a cheap procedure, for instance if Ω is given by either the nonnegative orthant or the Cartesian product of closed and bounded intervals, possibly together with a linear equality constraint. The choice of the scaling matrix D_k must keep the projection computationally nonexpensive: this issue will be discussed in Sect. 3.2.

3.1 Convergence

We recall the main convergence results for the SGP algorithm just now detailed: the first one, reported in Theorem 1, holds without the assumption of convexity for the functional to minimize f_0, while the second one (Theorem 2) is stronger but it can be proved only for convex objective functions.

Theorem 1 ([12, Theorem 2.1]) *Assume that f_0 in (21) is a differentiable function and the level set $X_0 = \{x \in \Omega : f_0(x) \leq f_0(x^{(0)})\}$ is bounded. Every accumulation point of the sequence $\{x^{(k)}\}_{k \in \mathbb{N}}$ generated by the SGP algorithm is a stationary point of (21).*

In [14] the convergence properties for SGP have been refined. The authors establish the criteria needed to guarantee the convergence of the sequence $\{x^{(k)}\}_{k \in \mathbb{N}}$ to a solution of (21) (see Theorem 2). Before introducing this result we recall the following definition.

Definition 1 Let $A, B \in \mathbb{R}^{n \times n}$ be symmetric and positive definite matrices. The notation $A \succeq B$ indicates that $A - B$ is a symmetric and positive semidefinite matrix or, equivalently, $x^T A x \geq x^T B x$ for $x \in \mathbb{R}^n$.

Theorem 2 ([14, Theorem 3.1]) *Assume that the objective function of* (21) *is convex and the solution set is not empty. Let* $\{x^{(k)}\}_{k\in\mathbb{N}}$ *be the sequence generated by SGP where* $D_k \in \mathcal{D}_\mu$, $\mu \geq 1$, *and satisfying the following condition*

$$(1 + \zeta_k)D_k \succeq D_{k+1} \quad \zeta_k \geq 0 \quad \sum_{k=0}^{+\infty} \zeta_k < \infty. \tag{24}$$

Then the sequence $\{x^{(k)}\}_{k\in\mathbb{N}}$ *converges to a solution of* (21).

Condition (24) states that the sequence $\{D_k\}_{k\in\mathbb{N}}$ asymptotically approaches a constant matrix [27, Lemma 2.3]. This requirement is not restrictive: it is easy to implement practical rules to fix matrices as in (24) as we will explain at the end of Sect. 3.2 (see Remark 1). The key point in selecting D_k is to consider matrices belonging to the compact set $\mathcal{D}_\mu$.

We emphasize that both the convergence theorems for SGP hold for any choice of the steplength α_k in the interval $[\alpha_{min}, \alpha_{max}]$ and the scaling matrix D_k in the set $\mathcal{D}_\mu$ since the sufficient decrease of the objective function is ensured by the Armijo-like backtracking procedure applied to select λ_k. For this reason, α_k and D_k can be considered as free parameters which can significantly optimize the convergence behaviour of the algorithm provided that they are chosen in a suitable way. Section 3.2 is devoted to clarify this aspect.

Together with the result shown in Theorem 2, in [14] the convergence rate with respect to the objective function values has been also provided. We resume the details in Theorem 3.

Theorem 3 ([14, Theorem 3.2]) *Assume that the objective function of* (21) *is convex and the solution set is not empty. In addition, suppose that* ∇f_0 *satisfies one of the following conditions:*

(a) ∇f_0 *is globally Lipschitz on* Ω;
(b) ∇f_0 *is locally Lipschitz and* f_0 *is level bounded on* Ω.

Let f_0^* *be the optimal function value for problem* (21). *Then, we have*

$$f_0(x^{(k)}) - f_0^* = \mathcal{O}(1/k).$$

We stress that, although the theoretical convergence rate estimate for SGP is only $\mathcal{O}(1/k)$, under suitable choices for α_k and D_k, the practical SGP behavior is very similar to a super-linear rate of convergence, especially in the first iterations, as shown by several numerical experiments in [14, 60].

3.2 Stepsize and Scaling Matrix Selection

The goodness of a first-order algorithm is strongly related to a proper selection of the steplength α_k. Therefore, its choice is a very crucial point and in the literature many efforts are known in this sense. Starting from the paper by Barzilai and Borwein (BB) [5], many BB-like steplength selection rules have been suggested to accelerate the performance of standard gradient methods [32–34, 36, 39, 40, 78]. Thanks to the nonrestrictive assumptions on α_k, any of these strategies can be adopted for the steplength in SGP. However, in [12], an updating scheme developed in [39] for nonscaled gradient methods and based on a proper alternation of the original BB rules has been generalized by modifying the standard BB rules in order to take into account the presence of the scaling matrix. These rules arise from the approximation of the Hessian $\nabla^2 f_0(x^{(k)})$ with the matrix $B(\alpha_k) = \alpha_k^{-1} D_k$ and by imposing the following quasi-Newton properties on $B(\alpha_k)$:

$$\alpha_k^{BB1} = \underset{\alpha_k \in \mathbb{R}}{\operatorname{argmin}} \| B(\alpha_k)s^{(k-1)} - z^{(k-1)} \|$$

$$\alpha_k^{BB2} = \underset{\alpha_k \in \mathbb{R}}{\operatorname{argmin}} \| s^{(k-1)} - B(\alpha_k)^{-1} z^{(k-1)} \|$$

where $s^{(k-1)} = x^{(k)} - x^{(k-1)}$ and $z^{(k-1)} = \nabla f_0(x^{(k)}) - \nabla f_0(x^{(k-1)})$. The resulting values become

$$\alpha_k^{BB1} = \frac{s^{(k-1)^T} D_k D_k s^{(k-1)}}{s^{(k-1)^T} D_k z^{(k-1)}}; \quad \alpha_k^{BB2} = \frac{s^{(k-1)^T} D_k^{-1} z^{(k-1)}}{z^{(k-1)^T} D_k^{-1} D_k^{-1} z^{(k-1)}};$$

which reduce to the standard BB rules when D_k is equal to the identity matrix for all k. Finally, the steplength selection strategy implemented within SGP consists in an adaptive alternation between the values

$$\bar{\alpha}_k = \max\{\alpha_{\min}, \min\{\alpha_{\max}, \alpha_k^{BB1}\}\}$$

and

$$\hat{\alpha}_k = \max\{\alpha_{\min}, \min\{\alpha_{\max}, \alpha_k^{BB2}\}\}$$

with $0 < \alpha_{\min} < \alpha_{\max}$.

As far as the scaling matrix is concerned, its selection usually must aim at two main goals: improving the convergence rate and avoiding to introduce significant computational costs. For these reasons, diagonal scaling matrices which add to the algorithm some local information of the optimization problem are typically considered. An example of this choice consists of taking into account a scaling matrix $D_k = \operatorname{diag}(d_1^{(k)}, d_2^{(k)}, ..., d_n^{(k)})$ which approximates the Hessian matrix $\nabla^2 f_0(x)$ by requiring

$$d_i^{(k)} = \left(\frac{\partial^2 f_0(x^{(k)})}{\partial^2 x_i} \right) \quad i = 1, ..., n$$

Since the computation of the Hessian could be very expensive, other possibilities should be analyzed. Particularly, the approach (13) based on the decomposition of ∇f_0 in a positive part and a non-negative one could be very fruitful since it is related to the nature of the problem to minimize and it can be always applied since such a decomposition always exists. Indeed, in [12], the authors suggest to select the elements $d_i^{(k)}$ of the diagonal scaling matrix D_k as

$$d_i^{(k)} = \max\left(\frac{1}{\mu}, \min\left(\mu, \frac{V_i(x^{(k)})}{x_i^{(k)}}\right)\right), \quad \mu \geq 1$$

in order to also guarantee that the eigenvalues of D_k lie in the compact set $[\frac{1}{\mu}, \mu]$.

However, the way to select other convenient scaling matrices is still an open problem.

Remark 1 We conclude the discussion about the selection of the scaling matrix by showing practical criteria in order to realize a sequence of scaling matrices fulfilling the requirement (24) needed to guarantee the convergence of the iterates generated by SGP to a solution of problem (21). More in detail, if the sequence $\{D_k\}_{k\in\mathbb{N}}$ is chosen according to

$$\{D_k\}_{k\in\mathbb{N}} \subset \mathcal{D}_{\mu_k} \quad \text{where} \quad \mu_k^2 = 1 + \zeta_k, \quad \zeta_k \geq 0, \quad \sum_{k=0}^{+\infty} \zeta_k < \infty, \qquad (25)$$

then condition (24) is satisfied. Indeed, since for any $D \in \mathcal{D}_\mu$, with $\mu > 1$, the following inequalities hold

$$\frac{1}{\mu}\|x\|^2 \leq x^T D x \leq \mu\|x\|^2, \quad \forall x \in \mathbb{R}^n$$

then, in view of (25),

$$x^T D_{k+1} x \leq \frac{\mu_{k+1}\mu_k}{\mu_k}\|x\|^2 \leq \mu_{k+1}\mu_k x^T D_k x \quad \forall x \in \mathbb{R}^n$$

which implies $D_{k+1} \preceq \mu_k\mu_{k+1} D_k$. Moreover $\mu_k\mu_{k+1}$ can be written as $\mu_k\mu_{k+1} = 1 + \xi_k$ where $\xi_k = \sqrt{(1 + \zeta_k)(1 + \zeta_{k+1})} - 1$. Since $\lim_{x\to 0}(\sqrt{1+x} - 1)/x = 1/2$ it follows that $\sum_{k=0}^{+\infty}\xi_k$ and $\sum_{k=0}^{+\infty}\zeta_k$ have the same behaviour. We can conclude that (25) implies (24). In practice, condition (25) ensures that the sequence $\{D_k\}_{k\in\mathbb{N}}$ approaches the identity matrix at a certain rate, while (24) implies the convergence to some symmetric positive definite matrix, as already mentioned.

4 A Variable Metric Inexact Linesearch Forward-Backward Method

In this section we come back to the original general problem (1). In [15], a variable metric inexact linesearch forward-backward algorithm (VMILA) of the form (4) has been proposed by following the key principle on which the SGP approach is based, namely, the possibility of freely select the parameters α_k and D_k thanks to a backtracking strategy along the descent directions which guarantees the convergence. To reach this goal, in [15] the authors generalize both the concept of descent direction by considering the proximal operator associated to the convex part of the objective function and the Armijo-like rule to determine the steplength along this direction ensuring the sufficient decrease of the objective function. To better detail this generalization process, we introduce the following preliminary notions.

4.1 Preliminary Notions

Definition 2 A vector $d \in \mathbb{R}^n$ is a descent direction for f in (1) at $x \in \text{dom}(f_1)$ if $f'(x; d) < 0$, where $f'(x; d)$ is the one-sided directional derivative of f at x with respect to d defined as

$$f'(x; d) = \lim_{\lambda \downarrow 0} \frac{f(x + \lambda d) - f(x)}{\lambda}$$

if the limit on the right-hand side exists in $\mathbb{R} \cup \{-\infty, +\infty\}$.

Definition 3 Let $\alpha \in \mathbb{R}$ be a positive parameter and $D \in \mathbb{R}^{n \times n}$ be a symmetric and positive definite matrix. Given $x \in \mathbb{R}^n$, we introduce the function $h_\alpha^D : \mathbb{R}^n \to \mathbb{R}$ defined as

$$h_\alpha^D(w; x) = \nabla f_0(x)^T (w - x) + \frac{1}{2\alpha}(w - x)^T D(w - x) + f_1(w) - f_1(x). \quad (26)$$

The function $h_\alpha^D(\cdot; x)$ is strongly convex and admits an unique minimum point for any $x \in \text{dom}(f_1)$. Moreover, the negative sign of h_α^D detects a descent direction, as stated in the following theorem.

Theorem 4 ([15, Proposition 2.2]) Let h_α^D be defined as in (26). If $x, z \in \text{dom}(f_1)$ and $h_\alpha^D(z; x) < 0$, then $f'(x, z - x) < 0$.

4.2 A Generalized Armijo Linesearch Along a Family of Descent Directions

By coming back to the scheme (4) and considering the notions just introduced, it is easy to prove that

$$y^{(k)} = \text{prox}_{\alpha_k f_1}^{D_k}(x^{(k)} - \alpha_k D_k^{-1} \nabla f_0(x^{(k)})) = \underset{w \in \mathbb{R}^n}{\text{argmin}} \, h_{\alpha_k}^{D_k}(w; x^{(k)}). \quad (27)$$

In addition, since $h_{\alpha_k}^{D_k}(x^{(k)}, x^{(k)}) = 0$ and $y^{(k)}$ is the minimum point of $h_{\alpha_k}^{D_k}(\cdot; x^{(k)})$, it holds that $h(y^{(k)}; x^{(k)}) < 0$, unless $y^{(k)}$ is a stationary point for (1). In view of Theorem 4, the vector $d^{(k)} = y^{(k)} - x^{(k)}$ is a descent direction for f at $x^{(k)}$. It is worth to remark that, actually, any vector $d^{(k)}$ which can be computed as $\tilde{y}^{(k)} - x^{(k)}$ with $\tilde{y}^{(k)} \in \text{dom}(f_1)$ and $h_{\alpha_k}^{D_k}(\tilde{y}^{(k)}; x^{(k)}) < 0$ is a descent direction.

Based on these considerations, a generalization of the monotone Armijo linesearch procedure (23) can be devised [15, 59, 74]. Suppose that a descent direction $d^{(k)} = \tilde{y}^{(k)} - x^{(k)}$, with $\tilde{y}^{(k)}, x^{(k)} \in \text{dom}(f_1)$, can be computed for the objective function at $x^{(k)}$. Given $\sigma, \beta \in (0, 1)$, the parameter λ_k is set equal to β^{m_k}, where m_k is the first non-negative integer m for which

$$f(x^{(k)}) - f(x^{(k)} + \beta^m d^{(k)}) \geq -\sigma \beta^m h_{\alpha_k}^{D_k}(\tilde{y}^{(k)}; x^{(k)}). \tag{28}$$

This backtracking strategy terminates in a finite number of steps if the two sequences $\{x^{(k)}\}_{k \in \mathbb{N}}$ and $\{\tilde{y}^{(k)}\}_{k \in \mathbb{N}}$ are such that

$$h_{\alpha_k}^{D_k}(\tilde{y}^{(k)}; x^{(k)}) < 0, \ \forall k. \tag{29}$$

Of course, if $\tilde{y}^{(k)} = y^{(k)}$ at each iteration, i.e., the minimization problem defining the proximal operator can be solved exactly, then the generalized Armijo linesearch is well defined. However, we stress that this is not needed: it is sufficient to find a convenient $\tilde{y}^{(k)}$ which verifies (29). This point allows to investigate how to inexactly compute $y^{(k)}$ in (4).

4.3 Inexact Computation of the Proximal Point

Let us start by recalling the following general result.

Theorem 5 ([15, Theorem 3.1]) *Let* $0 < \alpha_{min} < \alpha_{max}$ *and* $\mu > 1$ *and assume that* $\{\alpha_k\}_{k \in \mathbb{N}} \subset [\alpha_{min}, \alpha_{max}]$ *and* $\{D_k\}_{k \in \mathbb{N}} \subset \mathcal{D}_\mu$. *Let* $\{\tilde{y}^{(k)}\}_{k \in \mathbb{N}}$ *be a sequence of points in* $\text{dom}(f_1)$. *Let* $\{x^{(k)}\}_{k \in \mathbb{N}}$ *be the sequence generated by*

$$x^{(k+1)} = x^{(k)} + \lambda_k d^{(k)}, \quad d^{(k)} = \tilde{y}^{(k)} - x^{(k)},$$

where λ_k *is computed by means of* (28), $\tilde{y}^{(k)}$ *satisfies* (29) *and there exists* $K'' \subseteq \mathbb{N}$ *such that*

$$\lim_{k \in K'', k \to +\infty} h_{\alpha_k}^{D_k}(\tilde{y}^{(k)}; x^{(k)}) - h_{\alpha_k}^{D_k}(y^{(k)}; x^{(k)}) = 0, \quad \text{with } y^{(k)} = \underset{w \in \mathbb{R}^n}{\text{argmin}} \, h_{\alpha_k}^{D_k}(w; x^{(k)}). \tag{30}$$

Suppose that there exists a limit point $\bar{x}$ *of* $\{x^{(k)}\}_{k \in \mathbb{N}}$, *and let* $K' \supset K''$ *be a subset of indices such that* $\lim_{k \in K', k \to +\infty} x^{(k)} = \bar{x} \in \text{dom}(f_1)$. *Then* $\bar{x}$ *is a stationary point for problem* (1).

This theorem states that an approximation $\tilde{y}^{(k)}$ of $y^{(k)}$ has to verify both conditions (29) and (30) to ensure the stationarity of the limit points of $\{x^{(k)}\}_{k \in \mathbb{N}}$. In other words, $\tilde{y}^{(k)}$ has to guarantee that $d^{(k)}$ is a descent direction and, at the same time, to provide better and better approximations of $y^{(k)}$ as $k \to +\infty$. Unfortunately, requirement (30) can not be checked directly: for this reason, an implementable condition which implies (30) has been devised in [15, Sect. 3.2]. In particular, if the hypotheses of Theorem 5 hold but (30) is substituted by the following one

$$\frac{1}{\alpha_k} D_k(z^{(k)} - \tilde{y}^{(k)}) \in \partial_{\varepsilon_k} f_1(\tilde{y}^{(k)}) \tag{31}$$

where $z^{(k)} = x^{(k)} - \alpha_k D_k^{-1} \nabla f_0(x^{(k)})$ and $\{\varepsilon_k\}_{k \in \mathbb{N}} \subset \mathbb{R}_{\geq 0}$ such that $\lim_{k \to +\infty} \varepsilon_k = 0$, then the stationarity of the limit points of $\{x^{(k)}\}_{k \in \mathbb{N}}$ is still ensured. We observe that, thanks to the ε-subdifferential notion [66, Sect. 23], inclusion (31) can be viewed as a relaxation of the one which holds when the exact minimum point $y^{(k)}$ of $h_{\alpha_k}^{D_k}(\cdot; x^{(k)})$ can be computed, namely $\frac{1}{\alpha_k} D_k(z^{(k)} - y^{(k)}) \in \partial f_1(y^{(k)})$.

Remark 2 We discuss how to compute a point satisfying inclusion (31), for any given $\varepsilon_k \in \mathbb{R}_{\geq 0}$, when the convex function f_1 in (1) has the form

$$f_1(x) = g(Ax) + q(x) \tag{32}$$

where $A \in \mathbb{R}^{m \times n}$ and $g : \mathbb{R}^m \to \bar{R}$, $q : \mathbb{R}^n \to \bar{R}$ are convex, continuous functions whose proximal operators have closed-form expressions. The minimum problem (27) defining the proximity operator can be equivalently written in dual form as

$$\min_{y \in \mathbb{R}^n} h_{\alpha_k}^{D_k}(y; x^{(k)}) = \max_{v \in \mathbb{R}^m} \Psi_{\alpha_k}^{D_k}(v; x^{(k)})$$

$$= \max_{v \in \mathbb{R}^m} m^q(z^{(k)} - \alpha_k D_k^{-1} A^T v) - \frac{1}{2\alpha_k} \|z^{(k)} - \alpha_k D_k^{-1} A^T v\|_{D_k}^2 - g^*(v) + C_k$$

where $m^q(x) = \inf_{y \in \mathbb{R}^n} q(y) + \frac{1}{2\alpha_k} \|y - x\|_{D_k}^2$ is the Moreau envelope of parameters α_k, D_k associated to q, g^* is the Fenchel convex conjugate of g and C_k is a constant term independent from v. We remark that if $v^{(k)} = \operatorname{argmax}_{v \in \mathbb{R}^m} \Psi_{\alpha_k}^{D_k}(v; x^{(k)})$ then $y^{(k)} = \operatorname{prox}_{\alpha_k q}^{D_k}(z^{(k)} - \alpha_k D_k^{-1} A^T v^{(k)})$. For this reason, instead of computing the approximation $\tilde{y}^{(k)}$ of $y^{(k)}$ by means of a minimization iterative algorithm applied to (27), $\tilde{y}^{(k)}$ can be estimated by applying a maximization method to the dual problem. Particularly, if $\{v^{(k,l)}\}_{l \in \mathbb{N}}$ is the dual sequence converging to $v^{(k)}$, the corresponding primal sequence $\{\tilde{y}^{(k,l)}\}_{l \in \mathbb{N}}$ can be obtained as $\tilde{y}^{(k,l)} = \operatorname{prox}_{\alpha_k q}^{D_k}(z^{(k)} - \alpha_k D_k^{-1} A^T v^{(k,l)})$. The validity of (31) can be achieved by exploiting the following theorem which suggests a useful criterion to stop the dual iterative procedure.

Theorem 6 ([21, Proposition 4.2]) *Let $\varepsilon_k \in \mathbb{R}_{\geq 0}$. If*

$$h_{\alpha_k}^{D_k}(\tilde{y}^{(k)}; x^{(k)}) - \Psi_{\alpha_k}^{D_k}(v; x^{(k)}) \leq \varepsilon_k \tag{33}$$

with $\tilde{y}^{(k)} = prox_{\alpha_k q}^{D_k}(z^{(k)} - \alpha_k D_k^{-1} A^T v)$, *for some $v \in \mathbb{R}^m$, then (31) is satisfied.*

Thanks to Theorem 6, the iterative process employed to provide $\{v^{(k,l)}\}_{l \in \mathbb{N}}$ can be stopped when

$$h_{\alpha_k}^{D_k}(\tilde{y}^{(k,l)}; x^{(k)}) - \Psi_{\alpha_k}^{D_k}(v^{(k,l)}; x^{(k)}) \leq \varepsilon_k \qquad (34)$$

where the convergence of the sequence $\{\varepsilon_k\}_{k \in \mathbb{N}}$ to 0 is guaranteed by imposing, for instance, $\varepsilon_k = \frac{C}{k^p}$ with $C > 0$ and $p > 0$. Thus, the approximate computation of the proximal point through inequality (33) becomes automatically more accurate as the iterations proceed.

Finally, since the generalized Armijo backtracking strategy is well defined when $\tilde{y}^{(k)}$ also satisfies (29), condition (34) must be combined with $h_{\alpha_k}^{D_k}(\tilde{y}^{(k,l)}; x^{(k)}) < 0$.

4.4   Convergence

Theorem 5 states that all the limit points of the VMILA sequence are stationary for problem (1), provided that both the steplength α_k and the eigenvalues of D_k are chosen in prefixed positive intervals $[\alpha_{min}, \alpha_{max}]$ and $[\frac{1}{\mu}, \mu]$, respectively.

The next theorem investigates the conditions required to prove the convergence of the whole sequence of the iterates $\{x^{(k)}\}_{k \in \mathbb{N}}$ to a minimum point of (1).

Theorem 7 ([15, Theorem 3.3]) *Let $0 < \alpha_{min} < \alpha_{max}$ and $\mu \geq 1$. Assume that $\{\alpha_k\}_{k \in \mathbb{N}} \subset [\alpha_{min}, \alpha_{max}]$, the sequence $\{D_k\}_{k \in \mathbb{N}} \subset \mathcal{D}_\mu$ verifies (24), f_0 in (1) is convex and the solution set is not empty. Let $\{x^{(k)}\}_{k \in \mathbb{N}}$ be the sequence generated by*

$$x^{(k+1)} = x^{(k)} + \lambda_k d^{(k)}, \quad d^{(k)} = \tilde{y}^{(k)} - x^{(k)}, \qquad (35)$$

where λ_k is computed by means of (28) and $\tilde{y}^{(k)}$ satisfies (29) and (31) with a summable sequence $\{\varepsilon_k\}_{k \in \mathbb{N}}$. Then the sequence $\{x^{(k)}\}_{k \in \mathbb{N}}$ converges to a solution of (1).

With the next theorem we recall the convergence rate of VMILA.

Theorem 8 ([15, Theorem 3.5]) *Assume that the hypotheses of Theorem 7 hold and, in addition, that*

- *the sequence $\{\varepsilon_k\}_{k \in \mathbb{N}}$ satisfies $\varepsilon_k \leq -\tau\, h_{\alpha_k}^{D_k}(\tilde{y}^{(k)}; x^{(k)})$, $\tau > 0$;*
- *the gradient of f_0 is Lipschitz continuous on $dom(f_1)$.*

Let f^* be the optimal function value for problem (1). Then

$$f(x^{(k+1)}) - f^* = \mathcal{O}\left(\frac{1}{k}\right).$$

In [19] the authors proposed a slightly different version of the VMILA method: the new version (called VMILAn) has exactly the same features described in the previous sections for VMILA, but the $(k+1)-th$ iterate is updated as usual, namely $x^{(k+1)} = x^{(k)} + \lambda_k d^{(k)}$, if $f(x^{(k)} + \lambda_k d^{(k)}) < f(\tilde{y}^{(k)})$, otherwise $x^{(k+1)} = \tilde{y}^{(k)}$. For this modified VMILA version, in [19, Theorem 1], the authors prove the convergence of the iterates sequence by replacing the assumption of convexity for f_0 with the one that the objective function in (1) has to satisfy the Kurdyka-Łojasiewicz (K-L) property [19, Definition 3]. The details are provided in the following theorem.

Theorem 9 ([19, Theorem 1]) *Let $0 < \alpha_{\min} < \alpha_{\max}$ and $\mu \geq 1$. Assume that $\{\alpha_k\}_{k\in\mathbb{N}} \subset [\alpha_{\min}, \alpha_{\max}]$, $\{D_k\}_{k\in\mathbb{N}} \subset \mathcal{D}_\mu$, f in (1) is a K-L function, ∇f_0 is Lipschitz continuous and the solution set is not empty. Let $\{x^{(k)}\}_{k\in\mathbb{N}}$ be the sequence generated by*

$$x^{(k+1)} = \begin{cases} x^{(k)} + \lambda_k d^{(k)} & \text{if } f(x^{(k)} + \lambda_k d^{(k)}) < f(\tilde{y}^{(k)}) \\ \tilde{y}^{(k)} & \text{otherwise} \end{cases} \qquad (36)$$

where $d^{(k)} = \tilde{y}^{(k)} - x^{(k)}$, λ_k is computed by means of (28) and $\tilde{y}^{(k)}$ satisfies the following condition

$$\exists v^{(k)} \in \partial f(\tilde{y}^{(k)}) : \|v^{(k)}\| \leq b\|x^{(k+1)} - x^{(k)}\| + \vartheta_{k+1}, \quad \sum_{k=1}^{+\infty} \vartheta_k < \infty, \qquad (37)$$

for some $b > 0$, $\vartheta_k \in \mathbb{R}_{\geq 0}$. If $\{x^{(k)}\}_{k\in\mathbb{N}}$ admits a limit point $\overline{x}$, then the whole sequence converges to $\overline{x}$, which is stationary for problem (1).

The request that the functional to minimize fulfills the K-L property is not so restrictive; indeed examples of K-L functions are the indicator functions of semi-algebraic sets, real polynomials, p-norms and, in general, semi-algebraic functions or real analytic functions [11]. Moreover, when $\tilde{y}^{(k)} = y^{(k)}$, condition (37) is automatically guaranteed with $\vartheta_k \equiv 0$.

Remark 3 We observe that the conditions imposed on the sequences $\{\alpha_k\}_{k\in\mathbb{N}}$ and $\{D_k\}_{k\in\mathbb{N}}$ to guarantee the VMILA convergence are the same requested for the convergence of the SGP method. For this reason, all the considerations made for the selection of these parameters for SGP in Sect. 3.2 still hold for VMILA. In particular, we stress that α_k and D_k can be considered as "free" parameters which can be tuned for improving the algorithmic performances.

4.5 Generalizations

In the recent works [16, 20, 63], the proposed VMILA algorithm has been further generalized in order to either exploit the block-decomposable structure of certain optimization problems, and/or employ more general distances in the proximal operator (5). We now provide a brief overview of these admissible generalizations.

A Block Coordinate Extension

In many applications arising from image and signal processing, such as blind deconvolution [2, 61] and non-negative matrix factorization [51], one needs to solve problems of the form (1) in which the term f_1 is additively separable, namely

$$f_1(x) = \sum_{i=1}^{p} h_i(x_i)$$

where the functions h_i are proper, convex, lower semicontinuous, and the blocks of variables x_i are such that $x = (x_1, \ldots, x_p) \in \mathbb{R}^n$. In this setting, it is preferable to adopt an alternating minimization (or Gauss–Seidel) strategy [44], in which the objective function is cyclically minimized with respect to a single block of variables while the other ones are fixed, generating a sequence $\{x^{(k)}\}_{k \in \mathbb{N}}$ with $x^{(k)} = (x_1^{(k)}, \ldots, x_p^{(k)})$ and

$$x_i^{(k+1)} \in \operatorname*{argmin}_{u \in \mathbb{R}^{n_i}} f_0(x_1^{(k+1)}, \ldots, x_{i-1}^{(k+1)}, u, x_{i+1}^{(k)}, \ldots, x_p^{(k)}) + h_i(u), \quad i = 1, \ldots, p.$$

$$(38)$$

Rather than solving subproblem (38) exactly, which could be impractical and lead to nonconvergent sequences without some strict convexity assumptions [44], one can instead address its proximal-linearized version [11], i.e.

$$x_i^{(k+1)} = \operatorname{prox}_{\alpha_i^{(k)} h_i} \left(x_i^{(k)} - \alpha_i^{(k)} \nabla_i f_0(\tilde{x}_i^{(k)}) \right), \quad i = 1, \ldots, p \qquad (39)$$

where $\alpha_i^{(k)} > 0$ is a suitable steplength and $\nabla_i f_0(\tilde{x}_i^{(k)})$ denotes the partial gradient of f_0 with respect to x_i at the point $\tilde{x}_i^{(k)} = (x_1^{(k+1)}, \ldots, x_{i-1}^{(k+1)}, x_i^{(k)}, x_{i+1}^{(k)}, \ldots, x_p^{(k)})$.

In order to improve the practical performance of (39), a variable metric is usually introduced into the computation of the proximal operator, by either employing a majorization–minimization technique [25] or bounding the spectrum of the scaling matrix with the Lipschitz constant of the problem [38]. However, these strategies are not flexible and can be quite limiting in some cases of interest, e.g. when the Lipschitz constant is not known or assumes big values. By contrast, the authors in [20] propose to combine the block–coordinate strategy (39) with the VMILA algorithm which, as we have seen, allows the user to freely select the steplengths and the scaling matrices in bounded sets. At each outer iteration $k \in \mathbb{N}$, the proposed method first generates p sequences of inner iterates $\{x_i^{(k,\ell)}\}_{\ell=1,\ldots,L_i^{(k)}}, i = 1, \ldots, p$, by applying the VMILA algorithm $L_i^{(k)}$ times, i.e.

$$x_i^{(k,0)} = x_i^{(k)}$$
$$z_i^{(k,\ell)} = x_i^{(k,\ell)} - \alpha_i^{(k,\ell)}(D_i^{(k,\ell)})^{-1}\nabla_i f_0(\tilde{x}_i^{(k,\ell)})$$
$$\tilde{y}_i^{(k,\ell)} \approx_{\varepsilon_i^{(k,\ell)}} \text{prox}_{\alpha_i^{(k,\ell)}h_i}(z_i^{(k,\ell)}) \tag{40}$$
$$x_i^{(k,\ell+1)} = x_i^{(k,\ell)} + \lambda_i^{(k,\ell)}(\tilde{y}_i^{(k,\ell)} - x_i^{(k,\ell)}), \quad \ell = 0, 1, \ldots, L_i^{(k)} - 1$$

where the number of steps $L_i^{(k)}$ is bounded from above by an a priori fixed constant L_i, the point $\tilde{y}_i^{(k,\ell)}$ satisfies conditions (29)–(31) with $\alpha^{(k)} = \alpha_i^{(k,\ell)}$, $D_k = D_i^{(k,\ell)}$, $z^{(k)} = z_i^{(k,\ell)}$ and $\varepsilon_k = \varepsilon_i^{(k,\ell)}$, and $\lambda_i^{(k,\ell)}$ is computed by imposing the Armijo-like condition (28). Then, the next outer iterate is computed as $x^{(k+1)} = (x_1^{(k,L_1^{(k)})}, \ldots, x_p^{(k,L_p^{(k)})})$.

Specularly to the convergence analysis carried out in the previous sections, the stationarity of the limit points of the iterates sequences is proved for the scheme (40) without any additional assumption on the objective function, by only requiring that the errors $\varepsilon_i^{(k,\ell)}$ in the computation of the inexact proximal points $\tilde{y}_i^{(k,\ell)}$ converge to zero [20, Theorem 1]. Furthermore, convergence of the whole sequence to the limit point holds for a modified version of the method if the K-L property is satisfied and the proximal points are computed exactly, as we detail in the following theorem.

Theorem 10 ([20, Theorem 2]) *Let* $0 < \alpha_{min} < \alpha_{max}$, $\mu \geq 1$ *and* $L_1, \ldots, L_p \in \mathbb{Z}^+$. *For all* $k \in \mathbb{N}$, *let* $\{x_i^{(k,\ell)}\}_{\ell=1,\ldots,L_i^{(k)}}$, $i = 1, \ldots, p$, *be the sequences generated in* (40) *with* $\alpha_i^{(k,\ell)} \in [\alpha_{min}, \alpha_{max}]$, $D_i^{(k,\ell)} \in D_\mu$, $\tilde{y}_i^{(k,\ell)} = prox_{\alpha_i^{(k,\ell)}h_i}(z_i^{(k,\ell)})$ *and* $1 \leq L_i^{(k)} \leq L_i$. *For all* $i = 1, \ldots, p$, *choose* $\bar{\ell}_i^{(k)} \in \{0, \ldots, L_i^{(k)} - 1\}$ *and let* $\{x^{(k)}\}_{k\in\mathbb{N}}$ *be the sequence defined as follows*

$$x^{(k+1)} = \begin{cases} (x_1^{(k,L_1^{(k)})}, \ldots, x_p^{(k,L_p^{(k)})}) & \text{if } f(x_1^{(k,L_1^{(k)})}, \ldots, x_p^{(k,L_p^{(k)})}) \leq f(\tilde{y}_1^{(k,\bar{\ell}_1^{(k)})}, \ldots, \tilde{y}_p^{(k,\bar{\ell}_p^{(k)})}) \\ (\tilde{y}_1^{(k,\bar{\ell}_1^{(k)})}, \ldots, \tilde{y}_p^{(k,\bar{\ell}_p^{(k)})}) & \text{otherwise} \end{cases} . \tag{41}$$

Suppose that f *in* (1) *is a KL-function and that* ∇f_0 *is locally Lipschitz continuous on* Ω_0. *If* $\{x^{(k)}\}_{k\in\mathbb{N}}$ *admits a limit point* $\bar{x}$, *then the whole sequence converges to* $\bar{x}$, *which is stationary for problem* (1).

Remark 4 The modified step (41) is analogous to the modification (36) needed in VMILAn, and imposes a further descent condition which may be stronger, in principle, than the one imposed at each inner iteration in (40). We also remark that Theorem 10 requires only local (instead of global) Lipschitz continuity, which is an improvement with respect to Theorem 9 and allows to apply the proposed block-coordinate algorithm (as well as VMILAn) to specific problems where global Lipschitz continuity is denied, such as image restoration in presence of Poisson noise.

Remark 5 Choosing a suitable number of inner steps $L_i^{(k)}$ is crucial in order to make algorithm (40) effective. One can either set a constant number of inner iterations over the outer iterations, i.e. $L_i^{(k)} \equiv L_i$, as successfully done in some blind deconvolution problems in astronomy [61], or adopt an automatic stopping criterion based on the optimality of the inner iterate for the subproblem (38). We refer the reader to [20, Sect. 4] for more details on this issue.

Employing Bregman Distances

Forward-backward methods can be further generalized if one replaces the (possibly scaled) Euclidean distance in the proximal operator (5) with the more general concept of Bregman distance (see e.g. [6] and references therein). In particular, one can easily provide a Bregman version of the VMILA algorithm. Indeed, given $\alpha > 0$, $\varphi : \mathbb{R}^n \to \bar{\mathbb{R}}$ a strictly convex and differentiable function on $\text{int}(\text{dom}(\varphi))$ and its associated Bregman distance D_φ defined as

$$D_\varphi(x, y) = \varphi(x) - \varphi(y) - \nabla\varphi(y)^T(x - y) \qquad (42)$$

for all $x \in \text{dom}(\varphi)$, $y \in \text{int}(\text{dom}\varphi)$, then one can introduce the function

$$h_\alpha^\varphi(w; x) = \nabla f_0(x)^T(w - x) + \frac{1}{\alpha}D_\varphi(w, x) + f_1(w) - f_1(x) \qquad (43)$$

and define a Bregman linesearch based forward-backward algorithm given by

$$
\begin{aligned}
z^{(k)} &= x^{(k)} - \alpha_k \nabla f_0(x^{(k)}) \\
\tilde{y}^{(k)} &\approx_{\varepsilon_k} \text{prox}_{\alpha_k f_1}^\varphi(z^{(k)}) = \underset{w \in \mathbb{R}^n}{\text{argmin}}\ h_{\alpha_k}^\varphi(w; x^{(k)}) \\
x^{(k+1)} &= x^{(k)} + \lambda_k(\tilde{y}^{(k)} - x^{(k)}).
\end{aligned}
\qquad (44)
$$

Algorithm (44) is a variant of VMILA in which the function $h_{\alpha_k}^{D_k}$ has been replaced with $h_{\alpha_k}^\varphi$. Some of the theoretical results proved for the VMILA algorithm still hold for (44), including the well-definedness of the generalized Armijo linesearch (28) [15, Proposition 3.1], the stationarity of the limit points (see [15, Theorem 3.1], [63, Theorem 2]) and the practical procedure to compute an inexact proximal-gradient point satisfying (31) [63, Sect. 3]. Open problems remain the convergence of the iterates generated by (44) either when convexity or the K-L property hold.

The Bregman scheme (44) could be advantageous in order to better capture some second order information of the function f_0 in (1); for instance, when f_0 is the Kullback–Leibler functional defined in (17), a natural choice for the Bregman distance D_φ is the Kullback–Leibler itself, i.e.

$$D_\varphi(x, y) = \sum_{i=1}^n (x_i + bg) \log\left(\frac{x_i + bg}{y_i + bg}\right) + y_i - x_i.$$

Numerical experiments in [16, 63] confirm the effectiveness of this approach in the context of Poisson image reconstruction.

5 A Variable Metric Forward-Backward Method with Extrapolation

Another popular approach to solve problem (1) is the forward-backward method with extrapolation. In [17], the inertial method (6) has been generalized by introducing a variable metric induced by a sequence of symmetric and positive definite scaling matrices with bounded eigenvalues. Before introducing the resulting scheme we detail the additional assumptions which have to be satisfied by the function f_0 in (1). We suppose that f_0 is convex and has an L-Lipschitz continuous gradient on a nonempty, closed, convex set Y, where $\mathrm{dom}(f_1) \subseteq Y \subseteq \mathrm{dom}(f_0)$. The variable metric FB method with extrapolation devised in [17] can be written as

$$
\begin{aligned}
w^{(k)} &= \mathbb{P}_Y^{D_k}(x^{(k)} + \beta_k(x^{(k)} - x^{(k-1)})) \\
z^{(k)} &= w^{(k)} - \alpha_k D_k^{-1} \nabla f_0(w^k) \\
x^{(k+1)} &= \mathrm{prox}_{\alpha_k f_1}^{D_k}(z^{(k)})
\end{aligned}
\tag{45}
$$

where

$\mathbb{P}_Y^{D_k}(\cdot)$ denotes the projection operator onto the set Y with respect to the norm induced by D_k;

α_k is adaptively computed via a backtracking procedure which guarantees, starting from α_{k-1}, that

$$
f_0(x^{(k+1)}) \leq f_0(w^{(k)}) + \nabla f_0(w^{(k)})^T(x^{(k+1)} - w^{(k)}) + \frac{1}{2\alpha_k}\|x^{(k+1)} - w^{(k)}\|_{D_k}^2 ;
\tag{46}
$$

β_k has the form

$$
\beta_k = \frac{t_{k-1} - 1}{t_k}, \quad \beta_0 = 0
\tag{47}
$$

with $\{t_k\}_{k\in\mathbb{N}}$ satisfying the condition

$$
t_{k-1}^2 + t_k - t_k^2 \geq 0, \quad t_k \geq 1, \quad t_{-1} = 1, t_0 = 1;
\tag{48}
$$

$\{D_k\}_{k\in\mathbb{N}}$ has to be suitably chosen in the compact set $\mathcal{D}_\mu$, $\mu \geq 1$.

Remark 6 Algorithm (45) differs from (6) not only for the presence of a variable metric, but also for the D_k-induced projection of $w^{(k)}$ onto the set Y. Indeed, a drawback in the use of method (6) is that it may be unfeasible when $\mathrm{dom}(f_0)$ in (1) does not coincide with the whole space $\mathbb{R}^n$, since the point $w^{(k)}$ computed in (6) does not necessarily belong to $\mathrm{dom}(f_0)$. The projection operator $\mathbb{P}_Y^{D_k}(\cdot)$ assures that $w^{(k)}$ belongs to a subset of $\mathrm{dom}(f_0)$ where ∇f_0 exists and is a Lipschitz-continuous function.

Remark 7 The linesearch (46) is well defined, namely it terminates in a finite number of steps [17, Sect. 3].

Remark 8 An example of sequence $\{t_k\}_{k\in\mathbb{N}}$ and corresponding $\{\beta_k\}_{k\in\mathbb{N}}$ satisfying (47), (48) is the following one

$$t_k = \begin{cases} 1 & k = -1, 0 \\ \frac{a}{k+a} & k \geq 1 \end{cases} \qquad \beta_k = \begin{cases} 0 & k = 0 \\ \frac{k-1}{k+a} & k \geq 1 \end{cases} \tag{49}$$

with $a \geq 2$.

5.1 Inexact Computation of the Proximal Point

In [21], the authors introduce the possibility of inexactly computing the proximal point $x^{(k+1)}$ in (45). The approach followed to define a suitable approximation $\tilde{x}^{(k+1)}$ of $x^{(k+1)}$ is analogous to the one described in Sect. 4.3 for selecting $\tilde{y}^{(k)}$ in the VMILA method. In particular, condition (31) is replaced by the following more general inexactness criterion

$$0 \in \partial_{\varepsilon_k} h_{\alpha_k}^{D_k}(\tilde{x}^{(k+1)}; w^{(k)}) \quad \forall k > 0, \tag{50}$$

where, in this case, the function $h_{\alpha_k}^{D_k}(x, w)$ is defined as

$$h_{\alpha_k}^{D_k}(x, w) = \frac{1}{2\alpha_k}\|x - w + \alpha_k D_k^{-1}\nabla f_0(w)\|_{D_k}^2 + f_1(x).$$

As explained in [21, Remark 2.7], condition (50) is equivalent to say that there exist $\bar{\varepsilon}_k, \hat{\varepsilon}_k \geq 0$ with $\bar{\varepsilon}_k + \hat{\varepsilon}_k \leq \varepsilon_k$, and $e^{(k)} \in \mathbb{R}^n$ with $\|e^{(k)}\|_{D_k}^2 \leq 2\alpha_k\hat{\varepsilon}_k$, such that

$$\frac{1}{\alpha_k}D_k(z^{(k)} - e^{(k)} - \tilde{y}^{(k)}) \in \partial_{\bar{\varepsilon}_k} f_1(\tilde{y}^{(k)}). \tag{51}$$

The previous differential inclusion generalizes the one in (31), since it introduces an error on the computation of the gradient, controlled by the parameter $\hat{\varepsilon}_k$, in addition to the error on the proximal operator, measured by $\bar{\varepsilon}_k$. In order to ensure the desired convergence properties, the authors in [21] require that the sequences $\{k^2\bar{\varepsilon}_k\}_{k\in\mathbb{N}}$ and $\{k\sqrt{\hat{\varepsilon}_k}\}_{k\in\mathbb{N}}$ are both summable; sufficient choices for ε_k to guarantee these requirements are either $\varepsilon_k = \mathcal{O}(1/k^p)$ with $p > 4$ or $\varepsilon_k = \bar{\varepsilon}_k = \mathcal{O}(1/k^p)$ with $p > 3$ when no computational errors on the gradient are introduced.

By following the same arguments of Remark 2, in the special case of f_1 as in (32), $\tilde{x}^{(k+1)}$ can be computed by means of an iterative scheme applied to the dual problem of $\min_{x\in\mathbb{R}^n} h_{\alpha_k}^{D_k}(x, w^{(k)})$.

5.2   Convergence

The main convergence properties of the inexact variant of algorithm (45) are reported
in this section. Of course, the same results hold true when the proximal point is
computed exactly. In particular, Theorem 11 states the convergence of the whole
sequence of the iterates $\{x^{(k)}\}_{k\in\mathbb{N}}$ to a minimizer of f in (1) while Theorem 12
and Theorem 13 state the corresponding convergence rate in the objective function
values.

Theorem 11 ([21, Theorem 3.3]) *Assume that $\{t_k\}$ and $\{\beta_k\}$ are chosen as in (49)
with a > 2 and let $\{D_k\}_{k\in\mathbb{N}} \subset \mathcal{D}_\mu$ be a sequence of operators satisfying (24). Moreover, suppose that $x^{(k+1)} = \tilde{x}^{(k+1)}$, where $\tilde{x}^{(k+1)}$ fulfills (51) with $\{k^2\bar{\varepsilon}_k\}_{k\in\mathbb{N}}$ and
$\{k\sqrt{\hat{\varepsilon}_k}\}_{k\in\mathbb{N}}$ summable. Then, the sequence $\{x^{(k)}\}_{k\in\mathbb{N}}$ converges to a minimizer of
f in (1).*

Theorem 12 ([21, Theorem 3.1]) *Let $\{D_k\}_{k\in\mathbb{N}} \subset \mathcal{D}_\mu$ be a sequence of operators
satisfying (24) and assume that $\{t_k\}_{k\in\mathbb{N}}$, $\{\beta_k\}_{k\in\mathbb{N}}$ are chosen as in (49) with $a \geq 2$.
Moreover, suppose that $x^{(k+1)} = \tilde{x}^{(k+1)}$, where $\tilde{x}^{(k+1)}$ fulfills (51) with $\{k^2\bar{\varepsilon}_k\}_{k\in\mathbb{N}}$ and
$\{k\sqrt{\hat{\varepsilon}_k}\}_{k\in\mathbb{N}}$ summable. Let f^* be the optimal function value for problem (1). Then,
there exists a constant C such that*

$$f(x^{(k)}) - f^* \leq \frac{C}{(k+a)^2},$$

for all $k \geq 1$.

Theorem 13 ([21, Theorem 3.2]) *Let $\{D_k\}_{k\in\mathbb{N}} \subset \mathcal{D}_\mu$ be a sequence of operators
satisfying (24) and assume that $\{t_k\}_{k\in\mathbb{N}}$, $\{\beta_k\}_{k\in\mathbb{N}}$ are chosen as in (49) with a >
2. Moreover, suppose that $x^{(k+1)} = \tilde{x}^{(k+1)}$, where $\tilde{x}^{(k+1)}$ (51) with $\{k^2\bar{\varepsilon}_k\}_{k\in\mathbb{N}}$ and
$\{k\sqrt{\hat{\varepsilon}_k}\}_{k\in\mathbb{N}}$ summable. Let f^* be the optimal function value for problem (1). Then*

$$f(x^{(k)}) - f^* = o\left(\frac{1}{k^2}\right),$$

for all $k \geq 1$.

Remark 9 As for SGP and VMILA, the sequence $\{D_k\}_{k\in\mathbb{N}}$ has to be fixed in the
compact set $\mathcal{D}_\mu$, $\mu \geq 1$ and requirement (24) has to be verified. For this reason, the
criteria to select the scaling matrix D_k remain the ones detailed in Sect. 3.2.

6   Scaling Techniques for Proximal ε-Subgradient Methods

The scaling techniques described in the previous sections can be adapted also in the
case when both f_0 and f_1 in (1) are convex, lower semicontinuous but nondifferentiable. In these framework, assuming that an (approximate) subgradient of f_0 can be

easily computed, we focus on the following iteration

$$x^{(k+1)} = \text{prox}_{\alpha_k f_1}^{D_k}(x^{(k)} - \alpha_k D_k^{-1} u^{(k)}),\tag{52}$$

where $u^{(k)} \in \partial_{\varepsilon_k} f_0(x^{(k)})$. Clearly, rule (52) is formally similar to the FB iteration (2) where λ_k is set to one and $\nabla f_0(x)$ is replaced by any element of $\partial_{\varepsilon_k} f_0(x^{(k)})$.

The case $D_k = I$ is well studied in the literature, especially when f_1 reduces to the indicator function of a closed convex set; without being exhaustive, we mention for example [1, 29, 30, 45, 50, 54–56, 65, 73]. In general, the convergence is analyzed under suitable assumptions on the parameters ε_k and α_k. Typically, the error sequence ε_k is required to converge to zero, which means that the approximate subgradient $u^{(k)}$ of f_0 must be computed with an increasing accuracy as the iterations proceeds. On the other side, the meaning of the stepsize α_k in (52) is very different than the role played by the steplength parameters in the VMILA iteration. Indeed, even when $\varepsilon_k = 0$, moving along a negative subgradient does not necessarily produce a decrease on the objective function; hence, a line search procedure is not well defined in this context.

The convergence analysis of subgradient methods in the literature is often performed under the *Ermoliev* or *diminishing, divergent series* steplength rule, consisting in any choice of α_k obeying $\lim_{k\to\infty} \alpha_k = 0$, $\sum_{k=0}^{\infty} \alpha_k = \infty$ or under the *diminishing, divergent series, square summable* steplength rule, which requires $\sum_{k=0}^{\infty} \alpha_k = \infty$ and $\sum_{k=0}^{\infty} \alpha_k^2 < \infty$. In particular, the Ermoliev rule guarantees that the sequence $f(x^{(k)})$ converges to f^* and that the distance of the iterates from the solution set goes to zero, while the additional requirement that the sequence is square summable allows to prove the convergence of the iterates to a minimum point. In both cases, the practical realization of the subgradient algorithms requires the user to provide an entire sequence of stepsize parameters obeying the above mentioned requirements. An alternative, adaptive approach to the stepsize selection has been proposed in [10, 22, 42, 54] and consists in defining

$$\alpha_k = \frac{f(x^{(k)}) - f_k}{\|u^{(k)}\|^2} \quad \text{or} \quad \alpha_k = \frac{f(x^{(k)}) - f_k}{\max\{1, \|u^{(k)}\|^2\}},\tag{53}$$

where f_k is an adaptively computed estimate of the optimal value f^*. In this case, convergence of the objective function values and of the distance from the solution set to zero can be proved.

A more general analysis of ε-subgradient methods of the form (52), including a variable metric associated to the matrix D_k in combination with square summable or adaptive stepsize rules, is performed in [18]. The key assumption on the scaling matrices sequence to guarantee the convergence properties is the practical version (25) of the property employed to prove the convergence of the VMILA iterates in the convex case. We report below the statement of the main convergence result with diminishing, divergent series, summable square rule, whose proof can be found in [18]. Similar results are obtained also in [1, Lemma 1] for ε-subgradient projection methods, i.e. $D_k = I$, $f_1 = \iota_\Omega$, and in [30, Theorem 2.6] for the case $\varepsilon_k = 0$ and $D_k = I$.

Theorem 14 *Let $\{x^{(k)}\}$ be the sequence generated by iteration (52), where $u^{(k)} \in \partial_{\varepsilon_k} f_0(x^{(k)})$, for a given sequence $\{\varepsilon_k\}$ of non-negative scalars. Assume that there exist two positive constants ρ_u, ρ_w and a sequence $\{w^{(k)}\}$, $w^{(k)} \in \partial f_1(x^{(k)})$ such that $\|u^{(k)}\| \le \rho_u$ and $\|w^{(k)}\| \le \rho_w$ (subgradient boundedness). Assume that $\{D_k\}$ is chosen so that the condition (25) holds and that α_k and ε_k satisfy*

$$\lim_{k\to\infty} \varepsilon_k = 0, \quad \sum_{k=0}^{\infty} \varepsilon_k \alpha_k < \infty, \tag{54}$$

$$\sum_{k=0}^{\infty} \alpha_k = \infty, \quad \sum_{k=0}^{\infty} \alpha_k^2 < \infty. \tag{55}$$

Then, setting $f^* = \inf_{x \in \mathbb{R}^n} f(x)$ (possibly $f^* = -\infty$), we have

- $\liminf_{k\to\infty} f(x^{(k)}) = f^*$;
- *if $\{x^{(k)}\}$ is bounded, there exists a limit point of it belonging to the set of solutions X^* of (1);*
- *if X^* is not empty, the sequence $\{x^{(k)}\}$ converges to a solution of (1) and $\lim_{k\to\infty} f(x^{(k)}) = f^*$;*
- *if X^* is empty, the sequence $\{x^{(k)}\}$ is unbounded.*

When $X^* \ne \emptyset$, we are able to provide a convergence rate estimate for method (52) with the steplength rule (54), (55). From Lemma 2.3 in [18] (see also Theorem 2 in [1]), we have that, when the steplength is chosen as $\alpha_k = \mathcal{O}(\frac{1}{k})$, there exists a subsequence $\{x^{(k_\ell)}\}$ of $\{x^{(k)}\}$ such that $f(x^{(k_\ell)}) - f(x^*) \le \frac{1}{\log(k_\ell)}$. However, in spite of this poor theoretical estimate, with a suitable choice of the scaling matrices $\{D_k\}$, the method (52) shows a practical performance, definitely better than the one of the nonscaled version, as highlighted in the numerical experiments in [18].

Borrowing the ideas in [22, 42], the iteration (52) can be equipped also with an adaptive procedure to compute α_k as

$$\alpha_k = \frac{f(x^{(k)}) - f_k^{lev}}{\max(1, \|u^{(k)} + w^{(k)}\|_{D_k^{-1}})}, \tag{56}$$

where $w^{(k)} \in \partial f_1(x^{(k)})$ and f_k^{lev} is an estimate of the optimal value f^*. The updating procedure for f_k^{lev} is quite complicated: roughly speaking, f_k^{lev} is reduced when $f(x^{(k)})$ is sufficiently close to it and it is increased if after several iterates the function value is still too far from it (see [18, 54] for more details).

Under the assumption (25) on the scaling matrices sequence, this adaptive choice of α_k guarantees the convergence of $\{f(x^{(k)})\}$ to the optimal value and that the distance of the iterates from the solution set goes to zero [18, Theorem 4.1].

6.1 A Scaled Primal–Dual Hybrid Gradient Method

Scaling techniques can be introduced also in primal-dual methods applied to the following instance of the problem (1):

$$\min_{x \in \mathbb{R}^n} \psi(x) + g(Ax) + f_1(x), \tag{57}$$

where $A \in \mathbb{R}^{m \times n}$, $\psi(x)$, $g(x)$ and $f_1(x)$ are convex, proper, lower semicontinuous functions such that $\text{diam}(\text{dom}(g^*))$ is finite and $g^*(y)$ is the Fenchel dual of g. In particular, we consider the following Scaled Primal–Dual Hybrid Gradient (SPDHG) method:

$$y^{(k+1)} = \text{prox}_{\tau_k g^*}(y^{(k)} + \tau_k A x^{(k)}), \tag{58}$$

$$u^{(k)} = p^{(k)} + A^T y^{(k+1)}, \tag{59}$$

$$x^{(k+1)} = \text{prox}_{\alpha_k f_1}^{D_k}(u^{(k)} - \alpha_k D_k^{-1} u^{(k)}), \tag{60}$$

where $p^{(k)} \in \partial_{\nu_k} \psi(x^{(k)})$, for some $\nu_k \geq 0$, and $\{\tau_k\}$, $\{\alpha_k\}$ are the dual and primal steplength sequences respectively. Method (58)–(60) actually is a special case of the scaled FB ε-subgradient method (52), where $f_0 = \psi + g \circ A$. The key point of this interpretation is that $A^T y^{(k+1)}$ is an ε-subgradient of $g \circ A$ at $x^{(k)}$ as stated in [13, Lemma 1]. More precisely, $A^T y^{(k+1)} \in \partial_{\gamma_k}(g \circ A)(x^{(k)})$, where $\gamma_k = g(Ax^{(k)}) + g^*(y^{(k+1)}) - y^{(k+1)T} Ax^{(k)}$. Moreover, it can be shown that, under some reasonable assumption (e.g. when g is Lipschitz continuous on its domain), the error parameter γ_k is controlled by the inverse of the dual stepsize parameter τ_k.

Thus, recalling the additivity of the ε-subgradient, we can conclude that

$$u^{(k)} = p^{(k)} + A^T y^{(k+1)} \in \partial_{\varepsilon_k} f(x^{(k)}), \quad \varepsilon_k = \nu_k + \gamma_k. \tag{61}$$

In view of this remark, two different versions of the SPDHG method can be implemented: given a sequence of scaling matrices $\{D_k\}$ satisfying (25), in one case the sequences $\{\tau_k\}$, $\{\alpha_k\}$, $\{\nu_k\}$ are user provided and chosen so that (54), (55) are satisfied, while in the other case, only $\{\tau_k\}$ and $\{\nu_k\}$ must be provided and $\{\alpha_k\}$ is adaptively computed as in (56) (see [18, Corollaries 5.1, 5.2]).

As concerns as the choice of the scaling matrices D_k, the Split-Gradient idea described in Sect. 2 can be still exploited as an inspiration for designing a well performing variable metric for the iteration (52) and, as a special case of it, for the primal-dual method (58)–(60). In particular, in [18] a recursive procedure to compute a decomposition of the subgradient $u^{(k)} = V(x^{(k)}) - U(x^{(k)})$ with $V(x^{(k)}) > 0$ and $U(x^{(k)}) \geq 0$ is included in the SPDHG method (58)–(60), when applied to the deblurring of an image corrupted by Poisson noise via the TV regularization. Then, the matrix D_k^{-1} in (60) is defined as a diagonal matrix whose entries are the projection of $x_i^{(k)}/V_i(x^{(k)})$ onto the set $[1/\sqrt{1 + \zeta_k}, \sqrt{1 + \zeta_k}]$ (see Algorithm 2 in [18] for details). Numerical experiments show that this scaling strategy can be very effective to improve the convergence behaviour with respect to the nonscaled version of the same method, inducing a faster decrease of the objective function value through the iterates.

7 Conclusions and Perspectives

In this paper we have presented a variable metric approach for first-order methods aimed at minimizing the sum of a differentiable term and a convex one. First analysed in the seminal works [48, 49], the so-called split-gradient strategy relies upon the decomposition of the gradient of a differentiable function into the difference of a positive part and a non-negative one, and generates a sequence of diagonal positive definite scaling matrices which capture some second order information of the function at the current iterate. Such a technique can be easily adapted to several minimization techniques, such as linesearch based forward–backward methods [12, 15, 19, 21], inertial schemes [17, 20] and ε−subgradient methods [18], provided that it is combined either with a suitable adaptive procedure or a sufficiently fast decreasing steplength.

Future work could address the following issues:

- the adaptation of the proposed methods to problems where both the data-fidelity function and the regularization term are nonconvex, such as sparsity-based applications where the ℓ_0−norm is employed;
- the convergence of the iterates, assuming that the K-L property holds, when the proximal operator is approximately computed through condition (31);
- the modification of the inner routine for the inexact computation of the proximal operator employed in Sects. 4 and 5; in particular, one could investigate an algorithm where only the descent condition (29) is imposed, thus removing the stopping criterion (34) to ensure (31).

References

1. Alber, Ya.I., Iusem, A.N., Solodov, M.V.: On the projected subgradient method for nonsmooth convex optimization in a Hilbert space. Math. Prog. **81**, 23–35 (1998)
2. Ayers, G.R., Dainty, J.C.: Iterative blind deconvolution method and its applications. Opt. Lett. **13**(7), 547–549 (1988)
3. Attouch, H., Bolte, J., Svaiter, B.F.: Convergence of descent methods for semi-algebraic and tame problems: proximal algorithms, forward-backward splitting, and regularized Gauss-Seidel methods. Math. Program. **137**, 91–129 (2013)
4. Attouch, H., Peypouquet, J.: The rate of convergence of Nesterov's accelerated forward-backward method is actually faster than $1/k^2$. SIAM J. Optim. **26**, 1824–1834 (2016)
5. Barzilai, J., Borwein, J.M.: Two-point step size gradient methods. IMA J. Numer. Anal. **8**, 141–148 (1988)

6. Bauschke, H.H., Bolte, J., Teboulle, M.: A descent lemma beyond Lipschitz gradient continuity: first-order methods revisited and applications. Math. Oper. Res. **4**(1), 330–348 (2016)
7. Beck, A., Teboulle, M.: A fast iterative shrinkage-thresholding algorithm for linear inverse problems. SIAM J. Imaging Sci. **2**, 183–202 (2009)
8. Bertero, M., Boccacci, P., Desiderà, G., Vicidomini, G.: Image deblurring with Poisson data: from cells to galaxies. Inverse Probl. **25**, 123006 (2009)
9. Bertero, M, Boccacci, P., Ruggiero, V.: Inverse Imaging with Poisson Data, pp. 2053–2563. IOP Publishing, Bristol (2018)
10. Bertsekas, D.: Nonlinear Programming. Athena Scientific, Belmont (1999)
11. Bolte, J., Sabach, S., Teboulle, M.: Proximal alternating linearized minimization for nonconvex and nonsmooth problems. Math. Program. **146**(1–2), 459–494 (2014)
12. Bonettini, S., Zanella, R., Zanni, L.: A scaled gradient projection method for constrained image deblurring. Inverse Probl. **25**(1), 015002 (2009)
13. Bonettini, S., Ruggiero, V.: On the convergence of primal-dual hybrid gradient algorithms for total variation image restoration. J. Math. Imaging Vis. **44**, 236–253 (2012)
14. Bonettini, S., Prato, M.: New convergence results for the scaled gradient projection method. Inverse Probl. **31**, 095008 (2015)
15. Bonettini, S., Loris, I., Porta, F., Prato, M.: Variable metric inexact line-search based methods for nonsmooth optimization. SIAM J. Optim. **26**, 891–921 (2016)
16. Bonettini, S., Prato, M., Rebegoldi, S.: A cyclic block coordinate descent method with generalized gradient projections. Appl. Math. Comput. **286**, 288–300 (2016)
17. Bonettini, S., Porta, F., Ruggiero, V.: A variable metric forward-backward method with extrapolation. SIAM J. Sci. Comput. **38**(4), A2558–A2584 (2016)
18. Bonettini, S., Benfenati, A., Ruggiero, V.: Scaling techniques for ε-subgradient methods. SIAM J. Optim. **26**(3), 1741–1772 (2016)
19. Bonettini, S., Loris, I., Porta, F., Prato, M., Rebegoldi, S.: On the convergence of a linesearch based proximal-gradient method for nonconvex optimization. Inverse Probl. **33**(5), 055005 (2017)
20. Bonettini, S., Prato, M., Rebegoldi, S.: A block coordinate variable metric linesearch based proximal gradient method. Comput. Optim. Appl. **71**(1), 5–52 (2018)
21. Bonettini, S., Rebegoldi, S., Ruggiero, V.: Inertial variable metric techniques for the inexact forward-backward algorithm. SIAM J. Sci. Comput. **40**(5), A3180–A3210 (2018)
22. Brännlund, U., Kiwiel, K.C., Lindberg, P.O.: A descent proximal level bundle method for convex nondifferentiable optimization. Oper. Res. Lett. **17**, 121–126 (1995)
23. Chambolle, A., Dossal, Ch.: On the convergence of the iterates of the "Fast iterative shrinkage/thresholding algorithm". J. Optim. Theory Appl. **166**, 968–982 (2015)
24. Chouzenoux, E., Pesquet, J.C., Repetti, A.: Variable metric forward-backward algorithm for minimizing the sum of a differentiable function and a convex function. J. Optim. Theory Appl. **162**, 107–132 (2014)
25. Chouzenoux, E., Pesquet, J.C., Repetti, A.: A block coordinate variable metric forward-backward algorithm. J. Global Optim. **66**(3), 457–485 (2016)
26. Combettes, P., Pesquet, J.C.: Proximal splitting methods in signal processing. In: Bauschke, H.H., Burachik, R.S., Combettes, P.L., Elser, V., Luke, D.R., Wolkowicz, H. (eds.), Fixed-Point Algorithms for Inverse Problems in Science and Engineering, pp. 185–212. Springer, New York (2011); Optim. Appl. 49
27. Combettes, P., Vũ, B.: Variable metric quasi-Féjer monotonicity. Nonlinear Anal. **78**, 17–31 (2013)
28. Combettes, P., Vũ, B.: Variable metric forward-backward splitting with applications to monotone inclusions in duality. Optimization **63**, 1289–1318 (2014)
29. Correa, R., Lemaréchal, C.: Convergence of some algorithms for convex minimization. Math. Program. **62**, 261–275 (1993)
30. Cruz, J.Y.B.: On proximal subgradient splitting method for minimizing the sum of two nonsmooth convex functions. Set-Valued Var. Anal. **25**(2), 245–263 (2017)

31. Daube-Witherspoon, M.E., Muehllehner, G.: An iterative image space reconstruction algorithm suitable for volume ECT. IEEE Trans. Med. Imaging **5**, 61–66 (1986)
32. Dai, Y.H., Fletcher, R.: On the asymptotic behaviour of some new gradient methods. Math. Program. **103**, 541–559 (2005)
33. Dai, Y.H., Fletcher, R.: New algorithms for singly linearly constrained quadratic programming problems subject to lower and upper bounds. Math. Program. **106**, 403–421 (2006)
34. Dai, Y.H., Hager, W.H., Schittkowski, K., Zhang, H.: The cyclic Barzilai-Borwein method for unconstrained optimization. IMA J. Numer. Anal. **26**, 604–627 (2006)
35. De Pierro, A.R: On the convergence of the iterative image space reconstruction algorithm for volume ECT. IEEE Trans. Med. Imaging **6**, 174–175 (1986)
36. Di Serafino, D., Ruggiero, V., Toraldo, G., Zanni, L.: On the steplength selection in gradient methods for unconstrained optimization. Appl. Math. Comput. **318**, 176–195 (2018)
37. Drori, Y., Teboulle, M.: Performance of first-order methods for smooth convex minimization: a novel approach. Math. Program. **145**, 1–32 (2013)
38. Frankel, P., Garrigos, G., Peypouquet, J.: Splitting methods with variable metric for Kurdyka-Łojasiewicz functions and general convergence rates. J. Optim. Theory Appl. **165**, 874–900 (2015)
39. Frassoldati, G., Zanghirati, G., Zanni, L.: New adaptive stepsize selections in gradient methods. J. Ind. Manag. Optim. **4**(2), 299–312 (2008)
40. Friedlander, A., Mart, J.M., Molina, B., Raydan, M.: Gradient method with retards and generalizations. SIAM J. Numer. Anal. **36**, 275–289 (1999)
41. Galic, I., Weickert, J., Welk, M., Bruhn, A., Belyaev, A.G., Seidel, H.-P.: Image compression with anisotropic diffusion. J. Math. Imaging Vis. **31**(2–3), 255–269 (2008)
42. Goffin, J.L., Kiwiel, K.C.: Convergence of a simple subgradient level method. Math. Program. **85**, 207–211 (1999)
43. Grippo, L., Lampariello, F., Lucidi, S.: A nonmonotone line search technique for Newton's method. SIAM J. Numer. Anal. **23**(4), 707–716 (1986)
44. Grippo, L., Sciandrone, M.: On the convergence of the block nonlinear Gauss-Seidel method under convex constraints. Oper. Res. Lett. **26**(3), 127–136 (2000)
45. Kiwiel, K.C.: Convergence of approximate and incremental subgradient methods for convex optimization. SIAM J. Optim. **14**(3), 807–840 (2004)
46. Iusem, A.N.: Convergence analysis for a multiplicatively relaxed EM algorithm. Math. Methods Appl. Sci. **14**, 573–593 (1991)
47. Lange, K., Carson, R.: EM reconstruction algorithms for emission and transmission tomography. J. Comput. Assist. Tomogr. **8**, 306–316 (1984)
48. Lantéri, H., Roche, M., Cuevas, O., Aime, C.: A general method to devise maximum likelihood signal restoration multiplicative algorithms with non-negativity constraints. Signal Process. **81**(5), 945–974 (2001)
49. Lantéri, H., Roche, M., Aime, C.: Penalized maximum likelihood image restoration with positivity constraints: multiplicative algorithms. Inverse Probl. **18**, 1397–1419 (2002)
50. Larsson, T., Patriksson, M., Strömberg, A.-B.: On the convergence of conditional ε-subgradient methods for convex programs and convex-concave saddle-point problems. Eur. J. Oper. Res. **151**, 461–473 (2003)
51. Lin, C.J.: Projected gradient methods for nonnegative matrix factorization. Neural Comput. **19**(10), 2756–2779 (2007)
52. Lucy, L.B.: An iterative technique for the rectification of observed distributions. Astronom. J. **79**, 745–754 (1974)
53. Mülthei, H.N., Schorr, B.: On properties of the iterative maximum likelihood reconstruction method. Math. Methods Appl. Sci. **11**, 331–342 (1989)
54. Nedić, A., Bertsekas, D.P.: Incremental subgradient methods for nondifferentiable optimization. SIAM J. Optim. **12**, 109–138 (2001)
55. Nesterov, Y.: Introductory Lectures on Convex Optimization: A Basic Course. Applied Optimization, vol. 87. Kluwer Academic Publishers, Boston (2004)

56. Helou Neto, E.S., De Pierro, A.R.: Incremental subgradients for constrained convex optimization: a unified framework and new methods. Math. Program. **103**, 127–152 (2005)
57. Ochs, P., Chen, Y., Brox, T., Pock, T.: iPiano: Inertial proximal algorithm for non-convex optimization. SIAM J. Imaging Sci. **7**, 1388–1419 (2014)
58. Polyak, B.T.: Some methods of speeding up the convergence of iteration methods. USSR Comput. Math. Math. Phys. **4**, 1–17 (1964)
59. Porta, F., Loris, I.: On some steplength approaches for proximal algorithms. Appl. Math. Comput. **253**, 345–362 (2015)
60. Porta, F., Prato, M., Zanni, L.: A new steplength selection for scaled gradient methods with application to image deblurring. J. Sci. Comput. **65**, 895–919 (2015)
61. Prato, M., La Camera, A., Bonettini, S., Rebegoldi, S., Bertero, M., Boccacci, P.: A blind deconvolution method for ground based telescopes and Fizeau interferometers. New Astron. **40**, 1–13 (2015)
62. Raginsky, M., Willett, R.M., Harmany, Z.T., Marcia, R.F.: Compressed sensing performance bounds under Poisson noise. IEEE Trans. Signal Process. **58**, 3990–4002 (2010)
63. Rebegoldi, S., Bonettini, S., Prato, M.: A Bregman inexact linesearch-based forward-backward algorithm for nonsmooth nonconvex optimization. J. Phys. Conf. Ser. **1131**, 012013 (2018)
64. Richardson, W.H.: Bayesian-based iterative method of image restoration. J. Opt. Soc. Am. A **62**, 55–59 (1972)
65. Robinson, S.M.: Linear convergence of epsilon-subgradient descent methods for a class of convex functions. Math. Program. Ser. A **86**, 41–50 (1999)
66. Rockafellar, R.T.: Convex Analysis. Princeton University Press, Princeton (1970)
67. Salvo, K., Defrise, M.: A convergence proof of MLEM and MLEM-3 with fixed background. IEEE Trans. Med. Imaging. (2018). in press. https://doi.org/10.1109/TMI.2018.287096
68. Salzo, S., Villa, S.: Inexact and accelerated proximal point algorithms. J. Convex Anal. **19**, 1167–1192 (2012)
69. Salzo, S.: The variable metric forward-backward splitting algorithm under mild differentiability assumptions. SIAM J. Optim. **27**(4), 2153–2181 (2017)
70. Schmidt, M., Le Roux, N., Bach, F.: Convergence rates of inexact proximal-gradient methods for convex optimization. In: Proceedings of the 24th International Conference on Neural Information Processing Systems, pp. 1458–1466 (2011)
71. Sciacchitano, F., Dong, Y., Zeng, T.: Variational approach for restoring blurred images with Cauchy noise. SIAM J. Imaging Sci. **8**(3), 1894–1922 (2015)
72. Shepp, L.A., Vardi, Y.: Maximum likelihood reconstruction for emission tomography. Trans. Med. Imaging **1**, 113–122 (1982)
73. Shor, N.Z.: Minimization Methods for Nondifferentiable Functions. Springer, Berlin (1985)
74. Tseng, P., Yun, S.: A coordinate gradient descent method for nonsmooth separable minimization. Math. Program. **117**, 387–423 (2009)
75. Villa, S., Salzo, S., Baldassarre, L., Verri, A.: Accelerated and inexact forward-backward algorithms. SIAM J. Optim. **23**, 1607–1633 (2013)
76. Vogel, C.R.: Computational Methods for Inverse Problems. SIAM, Philadelphia (2002)
77. Zhang, J., Hu, Y., Nagy, J.G.: A scaled gradient method for digital tomographic image reconstruction. Inverse Probl. Imag. **18**, 239–259 (2018)
78. Zhou, B., Gao, L., Dai, Y.H.: Gradient methods with adaptive step-sizes. Comput. Optim. Appl. **35**(1), 69–86 (2006)

Structure Preserving Preconditioning for Frame-Based Image Deblurring

Davide Bianchi, Alessandro Buccini and Marco Donatelli

Abstract Regularizing preconditioners for accelerating the convergence of iterative regularization methods and improving their accuracy have been extensively investigated both in Hilbert and Banach spaces. For deconvolution problems, the classical approach defines preconditioners based on the circular convolution. On the other hand, for ℓ_2 regularization methods, it has been recently shown that a preconditioner preserving the structure of the convolution operator can be more effective. Such a preconditioner can improve both restoration quality and robustness of the method with respect to the choice of the regularization parameter when compared with the non-structured ones. In this paper we explore the use of structure preserving preconditioning for ℓ_1-norm regularization in the wavelet domain in image deblurring. A recently proposed preconditioned variant of the linearized Bregman iteration is modified to preserve the structure of the coefficient matrix according to the imposed boundary conditions. The structured preconditioner is chosen as an approximation of a regularized inverse of the convolution matrix. Selected numerical experiments show that our preconditioning strategy improves the previous results obtained with circulant preconditioning providing restorations with lower ringing effects and sharper details.

Keywords Image deblurring · Sparse regularization · Structured preconditioning

D. Bianchi (✉) · M. Donatelli
Dipartimento di Scienza e Alta Tecnologia, Università dell'Insubria, 22100 Como, Italy
e-mail: d.bianchi9@uninsubria.it

M. Donatelli
e-mail: marco.donatelli@uninsubria.it

A. Buccini
Dipartimento di Matematica e Informatica, Università di Cagliari, 09123 Cagliari, Italy
e-mail: alessandro.buccini@unica.it

© Springer Nature Switzerland AG 2019
M. Donatelli and S. Serra-Capizzano (eds.), *Computational Methods for Inverse Problems in Imaging*, Springer INdAM Series 36,
https://doi.org/10.1007/978-3-030-32882-5_2

1 Introduction

Image deblurring is the process of reconstructing an approximation of an image from blurred and noisy measurements. By assuming that the point spread function (PSF) is known, the observed image G is obtained from the convolution of the PSF with the true image F. The formation process of G, as well as the transmission process, produces some errors that we assume to be additive and with a Gaussian distribution. Therefore, denoting by $\mathbf{g}$ and $\mathbf{f}$ the stack ordered vectors corresponding to G and F, the discrete convolution problem is modeled by the linear system

$$\mathbf{g} = A\mathbf{f} + \boldsymbol{\eta}, \tag{1}$$

where A is the convolution matrix and $\boldsymbol{\eta}$ the noise vector. For the sake of simplicity we consider $n \times n$ images, thus $\mathbf{g}, \mathbf{f}, \boldsymbol{\eta}$ are vectors of $N = n^2$ components and the matrix A is $N \times N$ when proper Boundary Conditions (BCs) are imposed. BCs try to capture and include in the deblurring model the unknown behavior of the image outside the field of view in which the detection is made: see Sect. 3 and [29].

Thus, image deblurring consists in computing an approximation of the true image $\mathbf{f}$ by means of an appropriate solution of (1). Since the singular values of A gradually approach zero without a significant gap, independently of the BCs, A is very ill-conditioned and may be singular. Linear systems of equations with a matrix of this kind are commonly referred as linear discrete ill-posed problems and require regularization; see [26] for more details on discrete ill-posed problems. Therefore, a good approximation of $\mathbf{f}$ cannot be obtained from the algebraic solution (e.g., the least-square solution) of (1), but regularization methods are required.

The structure of the matrix A depends on the properties of the basic blurring model, i.e., the PSF and the BCs. In this work we assume that the blurring model is space-invariant, i.e., that it does not depend on the location, while the BCs can be defined by vary extrapolation strategies and are not necessarily defined as affine relations between the unknowns inside the field of view. For example, when periodic BCs are imposed, the matrix A is block circulant with circulant blocks (BCCB) and it is diagonalizable by discrete Fourier transform. For other BCs the matrix vector product with A can always be computed by means of the Fast Fourier Transform (FFT) on an appropriately padded image of larger size. On the other hand, the (pseudo) inverse of A cannot always be computed by fast trigonometric transforms, in particular when the PSF is not symmetric in both the horizontal and the vertical direction. Therefore, iterative methods are often preferable since the matrix A is never stored and only the PSF and the imposed BCs are necessary for the matrix-vector product.

As far as iterative regularization methods are concerned, these methods typically suffer of one of the following two shortcomings: either they are extremely slow like, e.g., the so-called Landweber iteration, or they are reasonably fast, but may deteriorate if not terminated appropriately. We refer to [22] for a comprehensive discussion of these and further properties of iterative regularization methods for linear ill-posed problems. Preconditioners can be used to accelerate the convergence, cf. [3, 23, 28,

30, 32], but an imprudent choice may spoil the quality of the computed restorations. Iterative regularization methods exhibit the so-called semiconvergence property: in the first iterations they reduce the algebraic error in the well-conditioned subspace of the low frequencies; but, when the algebraic error is reduced in the high frequencies space, the restoration error increases and the noise is amplified. Therefore, a fair estimation of the stopping iteration is crucial. Since the iterations are stopped before convergence, differently from the well-posed case, the choice of the preconditioner for iterative regularization methods do not only affect the speed of convergence, but also the reconstructed solution. In particular, a good preconditioner should not only accelerate the iteration, but has to lower the optimal reconstruction error achieved before the semiconvergence effect deteriorates the quality of the reconstructed solution.

To the best of our knowledge, the only preconditioners that preserve exactly the same structure of A have been proposed in [27] for zero Dirichlet BCs and symmetric PFS and in [15] for every BC and generic PSF. Note that the proposal in [15] is slightly different from those in [27] but they are asymptotically (in the size of the problem) equivalents.

In this paper, motivated by the fact that most real images usually have sparse approximations under some wavelet basis, we consider a regularization strategy based on tight frame decomposition that has been recently largely investigated [6–10]. In order to obtain a sparse approximation, we minimize the weighted ℓ_1-norm of the tight frame coefficients. Our aim is to explore the structure preserving preconditioning strategy proposed for a Landweber-like iteration in [15], in connection with the iterative methods for ℓ_1 regularization like the linearized Bregman algorithm (LBA) [7, 9] and the iterative shrinkage thresholding algorithm (ISTA) [12].

Let W^T be a tight-frame synthesis operator such that $W^T W = I$, the frame coefficients of the original image $\mathbf{f}$ are $\mathbf{x}$ such that

$$\mathbf{f} = W^T \mathbf{x}. \tag{2}$$

After reformulating the deblurring problem (1) in terms of frame coefficients

$$\min_{\mathbf{x}} \{\mu \|\mathbf{x}\|_1 + \|\mathbf{x}\|^2 : A W^T \mathbf{x} = \mathbf{g}\},$$

where $\| \cdot \|$ denotes the Euclidean 2-norm, a regularized solution can be obtained by the linearized Bregman splitting algorithm [35]. This method is known to converge very slowly for image deblurring problems, hence a preconditioning strategy is usually employed; such a preconditioned method is referred to as Modified Linearized Bregman Algorithm (MLBA) [9]. Combining the recent preconditioned regularization iteration for least-square ill-posed problems in [16] with the linearized Bregman algorithm, a new preconditioned iteration, similar to MLBA, was proposed in [8]. It is usually more robust and provides better restorations than MLBA. Since the introduction of a preconditioner requires the estimation of a second parameter, a nonstationary preconditioner inspired by [19] was also investigated.

The preconditioners for wavelet or tight frame algorithms previously investigated in the literature are based on BCCB approximations of A. In this paper, we investigate the use of the structured preconditioner proposed in [15] in the nonstationary iteration proposed in [8]. The preconditioner is defined as a regularized inverse having the same structure, and hence the same BCs, of the blurring matrix. The nonstationary parameter is computed at each step by a BCCB approximation of the preconditioner such that the involved nonlinear problem becomes separable and it can be easily solved in $O(n^2)$ complexity by Newton's method. Unfortunately, the structure of the preconditioner does not allow a straightforward extension of the convergence result in [8]. Nevertheless, numerical results show that the ringing effects in the restored images are largely reduced using our structure preserving preconditioner in connection with the linearized Bregman algorithm. For stopping the iterations and for estimating the preconditioner's parameter at each iteration, we assume that an estimation of $\delta = \|\boldsymbol{\eta}\|_2$ is available.

The paper is organized as follows. Section 2 describes the structure and some properties of the blurring matrix. The structured preconditioner is introduced in Sect. 3 with its nonstationary version and the automatic estimation of the related parameter. The structured preconditioner is combined with the nonstationary iteration for ℓ_1-norm framelet regularization in Sect. 4. Section 5 collects some numerical results and comparisons. Finally, Sect. 6 is devoted to concluding remarks.

2 Blurring Matrix and BCs

Let $H \in \mathbb{R}^{m \times m}$ be the PSF. We assume that the position of the PSF center is known and it is denoted by the index $(0, 0)$. Thus, H can be depicted as

$$H = [[h_{j_1, j_2}]_{j_1 = -m_{1,1}}^{m_{2,1}}]_{j_2 = -m_{1,2}}^{m_{2,2}},$$

where $m_{1,i} + m_{2,i} + 1 = m$, for $i = 1, 2$, and the indices are shifted according to the center of the PSF.

The pixels h_{j_1, j_2} of the PSF can be interpreted as the Fourier coefficients of a function. We will refer to this function as the symbol associated with H. In details, if n is odd and the PSF is obtained observing a white pixel on black background in the middle of the $n \times n$ image, then $m = n$ and the associated symbol is defined as

$$\phi(x_1, x_2) = \sum_{j_1, j_2 = -(n-1)/2}^{(n-1)/2} h_{j_1, j_2} e^{i(j_1 x_1 + j_2 x_2)}, \qquad i^2 = -1 \tag{3}$$

where the coefficients h_{j_1, j_2} far from the central coefficient $h_{0,0}$ are zero, since we assume that the PSF has compact support. The symbol (3) provides the spectral behavior of the matrix A independently of the BCs and, in particular, the eigenvalues

Table 1 Pad of the original image F obtained by imposing the classical BCs considered in [29], with $F_c = \text{fliplr}(F)$, $F_r = \text{flipud}(F)$, and $F_{rc} = \text{flipud}(\text{fliplr}(F))$, where $\text{fliplr}(\cdot)$ and $\text{flipud}(\cdot)$ are the MATLAB functions that perform the left-right and up-down flip, respectively

Zero	Periodic	Reflective
0 0 0	$F\ F\ F$	$F_{rc}\ F_r\ F_{rc}$
0 F **0**	$F\ F\ F$	$F_c\ F\ F_c$
0 0 0	$F\ F\ F$	$F_{rc}\ F_r\ F_{rc}$

of A are asymptotically distributed as a uniform sampling of ϕ up to few outliers; see [11, 21] for more details.

We now discuss some classical BCs and the structure of the resulting matrix A which, in turn, can be exploited to achieve fast computations. Common approaches force a functional dependency between the elements of F external to the FOV and those internal to this area. This has the effect of extending F outside of the FOV without adding any unknowns to the associated image deblurring problem. The use of different BCs can be motivated by additional information on the true image.

Table 1 summarizes the definition of zero, periodic, and reflective BCs; for a detailed description refer to [29]. For antireflective BCs see the review paper [21] and the original proposal in [33]. More sophisticated BCs like the synthetic BCs proposed in [25] or the higher order BCs in [14, 17] could be applied as well.

The matrix-vector product with the matrix A can always be performed by FFT resorting to a proper padding of the vector depending on the imposed BCs. Using for instance the MATLAB **padarray** function, the same padding of Table 1 can be applied to the $n \times n$ 2D array corresponding to the vector of size $N = n^2$. Then the $2n \times 2n$ padded image is convolved (by FFT) with H. Finally, the central $n \times n$ image (centered at the index $(0, 0)$) is the result of the matrix-vector product with the BCs applied for the padding. This is the implementation used in the MATLAB toolbox **RestoreTools** [31].

The matrix-vector product with the matrix A^H can be computed with a similar algorithm when zero Dirichlet or periodic BCs are imposed. Indeed, using the fact that the adjoint operator of the convolution is the correlation, which differs from the convolution only for a change of sign, a common approach is to rotate the PSF of 180 degrees and then to apply the previous padding strategy. This is for instance the implementation used in **RestoreTools**. Unfortunately, as proven in [18], this algorithm implements the matrix-vector product with a matrix A' (deblurring), which is the discretization of the correlation with the imposed BCs and, for generic BCs, A' is not necessarily equal to A^H.

Concerning the structure of the matrix A, when zero Dirichlet or periodic BCs are imposed, the matrix A has a block Toeplitz with Toeplitz blocks (BTTB) or BCCB structure, respectively. Moreover, using to standard notation, see e.g. [11], the matrix A can be denoted by $T_n(\phi)$ and $C_n(\phi)$, respectively, where ϕ is defined in (3). The matrix $C_n(\phi)$ can be diagonalized by the discrete Fourier transform

$$F_n = \frac{1}{\sqrt{n}} \left[e^{-\frac{2\pi i j_1 j_2}{n}} \right]_{j_1, j_2 = 0}^{n-1}.$$

Namely, $C_n(\phi) = F_n^H D_n(\phi) F_n$ where $D_n(\phi)$ is a diagonal matrix whose entries are the eigenvalues of $C_n(\phi)$ and are as a uniform sampling of ϕ, i.e.,

$$D_n(\phi) = \operatorname{diag}_{j_2 = 0, \ldots, n-1} \left(\phi \left(\frac{2\pi j_2}{n} \right) \right).$$

Imposing different BCs, the resulting shift-invariant structure of the matrix A can be denoted by

$$A = \mathcal{M}_n(\phi) = C_n(\phi) + R + E,$$

where R is a matrix of small rank and E is a matrix of small norm. Therefore, the symbol ϕ describes the spectral behavior of the matrix A independently of the imposed BCs.

3   Structure Preserving Preconditioning

In order to introduce the structure preserving preconditioner, we consider the following nonstationary iteration

$$\mathbf{f}_{k+1} = \mathbf{f}_k + Z_k(\mathbf{g} - A\mathbf{f}_k) = \mathbf{f}_k + Z_k \mathbf{r}_k, \tag{4}$$

where $\mathbf{r}_k = \mathbf{g} - A\mathbf{f}_k$ denotes the residual at step k and the matrix Z_k depends on A. For instance, $Z_k = \tau A^H$ leads to the Landweber method for $0 < \tau < 2/\|A^H A\|$.

The matrix Z_k can be used to speed up the convergence of the Landweber method, which is well-known to have a very slow convergence, without spoiling the quality of the computed solution. The iterated Tikhonov method [26] is the iteration (4) where

$$Z_k = A^H (AA^H + \alpha_k I)^{-1} \tag{5}$$

and it can also be interpreted as a preconditioned Landweber iteration [32]. The iteration proposed in [19] is a special instance of (4) with Z_k chosen as a BCCB matrix obtained replacing A with $C_n(\phi)$ in Eq. (5).

In [15], the algorithm proposed in [19] has been modified for preserving the structure of the matrix $A = \mathcal{M}_n(\phi)$, depending on the BCs. More in details, in [19], the authors consider a preconditioner of the form

$$Z_k = C_n(\phi)^H \left(C_n(\phi) C_n(\phi)^H + \alpha_k I \right)^{-1},$$

Thus Z_k is a BCCB matrix for every k. In [15] the authors, select as preconditioner a matrix Z_k of the form

$$Z_k = \mathcal{M}_n(\psi),$$

where the function ψ is a regularized approximation of $1/\phi$ related to the Tikhonov regularization and Z_k has the same structure of A. This is motivated by the fact that the class of matrices $\mathcal{M}_n(\cdot)$ shows a quasi-algebra structure, like BTTB matrices, such that it holds

$$\mathcal{M}_n(g)\mathcal{M}_n(f) \approx \mathcal{M}_n(gf);$$

see [1]. For a fixed size n, the symbol ψ reduces to the trigonometric polynomial

$$\psi(x_1, x_2) = \sum_{j_1, j_2 = -(n-1)/2}^{(n-1)/2} b_{j_1, j_2}^k e^{i(j_1 x_1 + j_2 x_2)}, \quad i^2 = -1 \tag{6}$$

where the coefficients $b_{i,j}^k$ can be computed imposing N interpolation conditions on the chosen regularized approximation of $1/\phi$. Let (j_1^o, j_2^o) be the index in the 2D array H of $h_{0,0}$, i.e., of the central coefficient of the PSF, using the Tikhonov approximation of $1/\phi$, i.e., by approximating $1/\phi$ by

$$1/\phi \approx \frac{\bar{\phi}}{|\phi|^2 + \alpha_k} = \psi,$$

the 2D array

$$B_k = \left[b_{j_1, j_2}^k \right]_{j_1, j_2 = -(n-1)/2}^{(n-1)/2}$$

can be computed by:

1. $C = \left[c_{j_1, j_2} \right] = \text{FFT2}(\text{circshift}(H, (-j_1^o + 1, -j_2^o + 1)));$
2. $v_{j_1, j_2}^k = \frac{\bar{c}_{j_1, j_2}}{|c_{j_1, j_2}|^2 + \alpha_k}$, for $j_1, j_2 = 0, \ldots, n-1$, $V_k = \left[v_{j_1, j_2}^k \right];$
3. $B_k = \text{IFFT2}(\text{circshift}(V_k, (j_1^o - 1, j_2^o - 1)));$

where the MATLAB function circshift performs a circular shift of a number of entries according to the second parameter.

The matrix Z_k is then obtained as $Z_k = \mathcal{M}_n(\psi)$, i.e., as the convolution matrix with convolutional kernel B_k and same BCs of A. We observe that, since both Z_k and A are convolution matrix with the same BCs they have the same structure.

A different approximation of $1/\phi$ can be simply obtained replacing the Tikhonov filter at step 2. with a different filter.

Remark 1 The selection of the Tikhonov filter at step 2. implies that

$$Z_k = \mathcal{M}_n(\psi) \approx A^H (AA^H + \alpha_k I)^{-1}.$$

In particular, imposing periodic BCs, it holds

$$Z_k = C_n(\psi) = C_n(\phi)^H \left(C_n(\phi) C_n(\phi)^H + \alpha_k I \right)^{-1} = C_n \left(\frac{\bar{\phi}}{|\phi|^2 + \alpha_k} \right), \quad (7)$$

where Z_k is the BCCB preconditioner used in [16, 19] and B_k is the 2D array containing the eigenvalues of Z_k.

In practice, the matrix Z_k is never allocated but only B_k is required. The matrix vector product with Z_k is then computed by the same algorithm used for computing the matrix-vector product with the matrix A simply replacing the PSF H with the vector B_k.

The choice of α_k is crucial for obtaining a good approximation within few iterations. A stationary choice $\alpha_k = \alpha$ requires a fair estimation of the parameter α. A too small α highly speeds up the convergence, but the method could become unstable, affecting the computed solution by noise amplification and hardly ever providing then a good estimation at the stopping iteration. On the other hand, a too large α does not necessarily accelerate the convergence.

Remark 2 For ill-posed problems, a regularizing preconditioner could slow down the convergence speed of the original iteration especially if it is defined for improving the quality of the computed solution. This happen for instance by preconditioning GMRES [20] and it could happen also with our preconditioner for some choices of α_k.

In order to avoid the estimation of α, a nonstationary strategy can be employed. A simple idea is to use a decreasing geometric sequence $\alpha_k = \alpha_0 \theta^k$, where for image deblurring α_0 and θ can be fixed as $\alpha_0 = 1$ and $\theta = 0.8$, see [19].

Here we adopt the strategy proposed in [15] and based on the algorithm in [19]. The parameter α_k is dynamically estimated at every iteration using $Z_k = C_n(\psi)$ as defined in (7) with a few steps of Newton's method. At step k the parameter α_k is determined by solving the nonlinear separable equation

$$q_k \|\mathbf{r}_k\| = \|\mathbf{r}_k - C_n(\phi) C_n(\psi) \mathbf{r}_k\| = \|\hat{\mathbf{r}}_k - D_n \hat{\mathbf{r}}_k\|, \quad (8)$$

where $\hat{\mathbf{r}}_k = F_n^H \mathbf{r}_k$, $D_n = \mathrm{diag}_{j_1, j_2 = 0, \ldots, n-1} (|c_{j_1, j_2}|^2 / (|c_{j_1, j_2}|^2 + \alpha_k))$ and

$$q_k = \max \{ q, \ 2\rho + (1 + \rho \delta / \|r_k\|) \}.$$

The parameter $q = 0.7$ is included as a safeguard to prevent that q_k decreases too rapidly and hence the preconditioner deteriorates the quality of the computed solution.

Remark 3 The parameter ρ measures how much we trust in the approximation of our preconditioner and in the numerical results it is simply fixed as $\rho = 10^{-4}$.

Note that the choices of q and ρ agree with the choices in [8].

Assuming that

$$\|(C_n(\phi) - A)\mathbf{z}\| \leq \rho \|A\mathbf{z}\|, \qquad \forall \mathbf{z} \in \mathbb{R}^N, \tag{9}$$

where $0 < \rho < 1/2$, in [19] it is proved that, choosing α_k by the equation (8) and $Z_k = C_n(\psi)$ as in (7), the iteration (4) converges monotonically and it defines a regularization method. Imposing to Z_k the same structure as A, then condition (9) has no longer meaning and the convergence of the iteration (4) cannot be easily derived.

4 Preconditioned Iteration for ℓ_1-Norm Framelet Regularization

It is well known that many images have a sparse representation in the wavelet domain. A sparse representation in the computed solution can be enforced imposing the ℓ_1-norm of the wavelet coefficients in the regularization term.

The synthesis approach [9, 24] consists of solving the problem (1) for the framelet coefficient instead that for the image itself. Let $W^H \in \mathbb{R}^{n^2 \times s}$ with $s \geq n^2$ be a tight-frame or wavelet synthesis operator. Recall that $W^H W = I$, then we can rewrite (1) as

$$\mathbf{g} = AW^H W \mathbf{f} + \boldsymbol{\eta}.$$

Denoting by $\mathbf{x}$ the vector $W\mathbf{f}$ according to (2) yields

$$\mathbf{g} = AW^H \mathbf{x} + \boldsymbol{\eta}.$$

Thus, the coefficient matrix of the problem becomes

$$AW^H \in \mathbb{R}^{n^2 \times s},$$

Note that using tight-frames $W^H W = I$ but $WW^H \neq I$ [13]. The use of tight-frames instead of wavelets is motivated by the fact that the redundancy of tight-frame systems leads to robust signal representations in which partial loss of the data can be tolerated, without adverse effects, see e.g. [10];

Let the nonlinear soft-thresholding operator $\mathbf{S}_\mu$ be defined component-wise as

$$[\mathbf{S}_\mu(\mathbf{x})]_i = S_\mu(x_i) = \text{sgn}(x_i) \max\{|x_i| - \mu, 0\}. \tag{10}$$

The algorithm proposed in [8] can be expressed as

$$\begin{cases} \mathbf{t}_{k+1} = \mathbf{t}_k + WC_n(\phi)^H \left(C_n(\phi)C_n(\phi)^H + \alpha_k I\right)^{-1} (\mathbf{g} - AW^H \mathbf{x}_k), \\ \mathbf{x}_{k+1} = \mathbf{S}_\mu(\mathbf{t}_{k+1}). \end{cases} \tag{11}$$

We recall that, in [8], the condition (9) as well as the condition

$$\|C_n(\phi)W^H(\mathbf{u} - S_\mu(\mathbf{u}))\| \leq \rho\delta, \qquad \forall \mathbf{u} \in \mathbb{R}^s, \tag{12}$$

where $\delta = \|\boldsymbol{\eta}\|$ is the noise level and ρ comes from Remark 3 are assumed to be true. Assumption (12) is equivalent to consider the soft-threshold parameter μ as a continuous function with respect to the noise level δ, i.e., $\mu = \mu(\delta)$, and such that $\mu(\delta) \to 0$ as $\delta \to 0$. This is a common request in many soft-thresholding based methods [12].

In [8] is proven that if α_k is chosen by (8), then the iteration (11) terminates after $k = k_\delta \geq 0$ iterations with

$$\|\mathbf{r}_{k_\delta}\| \leq \tau\delta < \|\mathbf{r}_k\|, \qquad k = 0, 1, \cdots, k_\delta - 1, \tag{13}$$

where $\tau = (1 + 2\rho)/(1 - 2\rho)$. This stopping criterion is known as *discrepancy principle* [22], Moreover, assuming that $\mathbf{t}_0$ is not a solution of the linear system

$$\mathbf{g} = AW^H\mathbf{x}, \tag{14}$$

and that δ_m is a sequence of positive real numbers such that $\delta_m \to 0$ as $m \to \infty$, in [8] is proven that the sequence $\{\mathbf{x}_{k(\delta_m)}\}_{m\in\mathbb{N}}$, generated by the discrepancy principle (13) and algorithm (11), converges as $m \to \infty$ to the solution of (14) which is closest to $\mathbf{t}_0$ in Euclidean norm. We highlighted the dependency of $\mathbf{x}_{k(\delta_m)}$ to the noise level δ_m.

Our algorithm combines the structure preserving preconditioner described in the previous section with the algorithm (11). In detail, the factor

$$C_n(\phi)^H \left(C_n(\phi)C_n(\phi)^H + \alpha_k I\right)^{-1}$$

is replaced with $Z_k = \mathcal{M}_n(\psi)$ as done in [15] for the ℓ_2-norm iteration in (4). The resulting algorithm is

$$\begin{cases} \mathbf{t}_{k+1} = \mathbf{t}_k + W\mathcal{M}_n(\psi)(\mathbf{g} - AW^H\mathbf{x}_k), \\ \mathbf{x}_{k+1} = \mathbf{S}_\mu(\mathbf{t}_{k+1}), \end{cases} \tag{15}$$

where ψ is defined as in (6) and whose Fourier coefficients are computed by the procedure at points 1–3 in Sect. 3. The parameter α_k is estimated solving the usual Eq. (8) for circulant matrices for a certain ρ, which is fixed as $\rho = 10^{-4}$ in the numerical results. Observe that, as we pointed out in Sect. 3, the preconditioner Z_k has the same structure as the matrix A.

Unfortunately, like for the ℓ_2-norm regularization, when Z_k has a generic structure, the theoretical results in [8] cannot be easily extended. Nevertheless, the numerical results in the next section show that the quality of the computed solutions with our algorithm (15) is better compared to the algorithm (11). In particular, our algorithm presents reduced ringing effects at the boundary.

Since our algorithm (15) shows the semiconvergence property typical of iterative regularization methods, it is stopped according to the discrepancy principle (13), where we fix $\tau = 1.01$ in the numerical results. Note that $\tau = 1.01$ is a common choice for iterative regularization since the theoretical analysis requires $\tau > 1$ [22].

5 Numerical Results

In this section, we will show the numerical results for image deblurring comparing our proposal with other classical algorithms for sparse wavelet restoration like ISTA [12] and FISTA [2]. Moreover, we compare with the approach presented in [19] and with one of the extensions proposed in [4].

The regularization parameter μ in (10) is chosen by hand in order to provide the best restoration. An analysis of the role of the parameter μ in the Bregman algorithm can be found in [5].

The tight-frame used in our tests is the piecewise linear B-spline framelets given in [9]. Namely, given the masks

$$w_0 = \frac{1}{4}[1, \ 2, \ 1], \quad w_1 = \frac{\sqrt{2}}{4}[1, \ 0, \ -1], \quad w_2 = \frac{1}{4}[-1, \ 2, \ -1],$$

we define the 1D filters of size $n \times n$ by imposing reflective BCs

$$W_0 = \frac{1}{4}\begin{bmatrix} 3 & 1 & 0 & \dots & 0 \\ 1 & 2 & 1 & & \\ & \ddots & \ddots & \ddots & \\ & & 1 & 2 & 1 \\ 0 & \dots & 0 & 1 & 3 \end{bmatrix}, \quad W_1 = \frac{1}{4}\begin{bmatrix} 1 & -1 & 0 & \dots & 0 \\ -1 & 2 & -1 & & \\ & \ddots & \ddots & \ddots & \\ & & -1 & 2 & -1 \\ 0 & \dots & 0 & -1 & 1 \end{bmatrix},$$

and

$$W_2 = \frac{1}{4}\begin{bmatrix} -1 & 1 & 0 & \dots & 0 \\ -1 & 0 & 1 & & \\ & \ddots & \ddots & \ddots & \\ & & -1 & 0 & 1 \\ 0 & \dots & 0 & -1 & 1 \end{bmatrix}.$$

The nine 2D filters are obtained by

$$W_{i,j} = W_i \otimes W_j, \quad i, j = 0, 1, 2,$$

where $\otimes$ denotes the tensor product operator. Finally, the corresponding tight-frame analysis operator is

$$W = \begin{bmatrix} W_{0,0} \\ W_{0,1} \\ \vdots \\ W_{2,2} \end{bmatrix}.$$

Throughout the experiments, we set the level of the framelet decomposition to 3.

In all examples we impose reflective boundary conditions and we deblur the image using our proposed approach (NMLBA-Str), the original non-structured version (NMLBA) [8], AIT [19], AIT-GP [4], ISTA [12], and FISTA [2]. In order to compare these methods we consider three quantities, the relative restoration error (RRE) defined by

$$\text{RRE}(\mathbf{f}) = \frac{\|\mathbf{f} - \mathbf{f}_{\text{true}}\|}{\|\mathbf{f}_{\text{true}}\|},$$

where $\mathbf{f}_{\text{true}}$ denotes the exact solution of the problem, the peak signal to noise ratio (PSNR) defined by

$$\text{PSNR}(x) = 20 \log_{10} \left(\frac{255n^2}{\|\mathbf{f} - \mathbf{f}_{\text{true}}\|} \right),$$

and the structure similarity index (SSIM) defined in [34]. The definition of the SSIM is more involved, here we just recall that SSIM measures how well the overall structure of the image is recovered and that the higher the index the better the reconstruction. In particular, the highest value achievable is 1.

We set all the free parameters in the considered method so that the obtained reconstruction minimizes the RRE, or, equivalently, maximizes the PSNR. The ISTA and FISTA algorithm converges to the solution of the ℓ_2-ℓ_1 problem, thus we stop them when two consecutive iterations are close enough, i.e., when

$$\frac{\|\mathbf{x}_{k+1} - \mathbf{x}_k\|}{\|\mathbf{x}_k\|} < 10^{-4}.$$

Cameraman We consider the following image deblurring problem. We blur the cameraman image with a non-symmetric PSF and add 1% of white Gaussian noise; see Fig. 1.

We report the result obtained with the different methods in Table 2. From these results we can observe that our proposal is able to outperform all the other considered methods in accuracy and SSIM. This is confirmed by the visual inspection of the reconstructions in Figs. 2 and 3. We can observe that the proposed approach is able to reduce the artifact on the boundaries of the image. In particular, we can observe that the reconstruction obtained with both NMLBA and FISTA are affected by ringing around the edges. This ringing effect disappears when the structured preconditioner is applied. Moreover, the reconstruction obtained with FISTA presents many artifacts

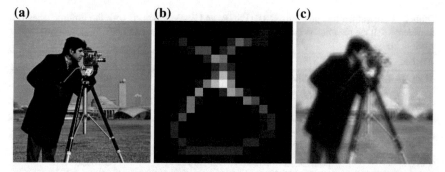

Fig. 1 Cameraman test problem: **a** True image $(238 \times 238$ pixels), **b** PSF $(17 \times 17$ pixels), **c** blurred image corrupted by 1% white Gaussian noise

Fig. 2 Cameraman test problem reconstructions obtained with different methods: **a** NMLBA-Str, **b** NMLBA, **c** FISTA

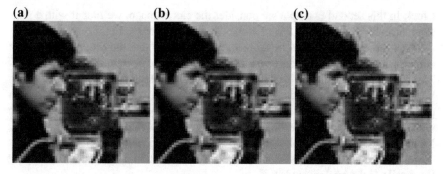

Fig. 3 Cameraman test problem blown-ups of the reconstructions obtained with different methods: **a** NMLBA-Str, **b** NMLBA, **c** FISTA

which result in the loss of details (see Fig. 3c), while the proposed approach is able to provide a much more accurate reconstruction of the details of the image as shown in Fig. 3a.

Table 2 Cameraman test problem: RRE, PSNR, and SSIM obtained with the different methods considered. All the parameters are tuned to obtain the highest value of PSNR. We highlight the best result in boldface

Method	RRE	PSNR	SSIM
NMLBA-Str	**0.068047**	**29.032**	**0.86794**
NMLBA	0.069711	28.822	0.86105
AIT	0.094712	26.160	0.73918
AIT-GP	0.089522	26.650	0.78689
ISTA	0.074759	28.215	0.83083
FISTA	0.073215	28.396	0.78956

(a)   **(b)**   **(c)**

Fig. 4 Clock test problem: **a** True image (235 × 235 pixels), **b** PSF (11 × 11 pixels), **c** blurred image corrupted by 2% white Gaussian noise

Clock In this second example we consider the clock image, we blur it with a non-symmetric PSF and we add 2% of white Gaussian noise; see Fig. 4.

From the results reported in Table 3 we can observe that the proposed approach outperforms the other considered method both in term of error and in term of SSIM. This is confirmed from the visual inspection of the reconstructions proposed in Fig. 5. We can observe that the structured preconditioner is able to dampen the ringing effect present in the reconstruction, especially around the boundary of the image and to reduce the presence of artifacts. This is evident in the blown-ups reported in Fig. 3. Moreover, we can observe that the proposed approach is able to denoise the image more effectively without oversmoothing the reconstruction, i.e., without destroying the details of the image; see Fig. 6.

6 Conclusions

We have combined the structured preconditioning proposed in [15] with the frame based iteration proposed in [8] for image deblurring problems. The resulting algorithm provides restorations with reduced ringing effects with respect to the algorithm in [8] and to classical wavelets based algorithms [2, 12, 24].

Table 3 Clock test problem: RRE, PSNR, and SSIM obtained with the different methods considered. All the parameters are tuned to obtain the highest value of PSNR. We highlight the best result in boldface

Method	RRE	PSNR	SSIM
NMLBA-Str	**0.054531**	**27.623**	**0.85855**
NMLBA	0.068787	25.605	0.81271
AIT	0.071323	25.291	0.68975
AIT-GP	0.066857	25.853	0.78177
ISTA	0.054695	27.597	0.77300
FISTA	0.063126	26.351	0.76676

Fig. 5 Clock test problem reconstructions obtained with different methods: **a** NMLBA-Str, **b** NMLBA, **c** ISTA

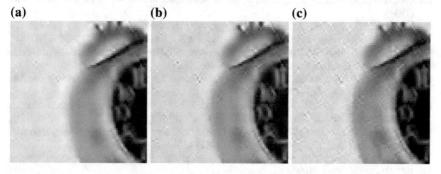

Fig. 6 Clock test problem blown-ups of the reconstructions obtained with different methods: **a** NMLBA-Str, **b** NMLBA, **c** ISTA

The goodness of our algorithm confirms that improvements in ℓ_2-norm regularization methods can be successfully applied to the inner steps of many nonlinear models.

Acknowledgements The authors are members of the INdAM Research group GNCS, which has partially supported this work.

References

1. Aricò, A., Donatelli, M., Serra-Capizzano, S.: Spectral analysis of the anti-reflective algebra. Linear Algebra Appl. **428**, 657–675 (2008)
2. Beck, A., Teboulle, M.: A fast iterative shrinkage-thresholding algorithm for linear inverse problems. SIAM J. Imaging Sci. **2**(1), 183–202 (2009)
3. Brianzi, P., Di Benedetto, F., Estatico, C.: Improvement of space-invariant image deblurring by preconditioned Landweber iterations. SIAM J. Sci. Comput. **30**, 1430–1458 (2008)
4. Buccini, A.: Regularizing preconditioners by non-stationary iterated Tikhonov with general penalty term. Appl. Num. Math. **116**, 64–81 (2017)
5. Buccini, A., Park, Y., Reichel, L.: Numerical aspects of the nonstationary modified linearized Bregman algorithm. Appl. Math. Comput. **337**(15), 386–398 (2018)
6. Buccini, A., Reichel, L.: An ℓ_2-ℓ_q regularization method for large discrete ill-posed problems. J. Sci. Comput., in Press
7. Cai, J.-F., Chan, R., Shen, L., Shen, Z.: Wavelet algorithms for high-resolution image reconstruction. SIAM J. Sci. Comput. **24**, 1408–1432 (2003)
8. Cai, Y., Donatelli, M., Bianchi, D., Huang, T.Z.: Regularization preconditioners for frame-based image deblurring with reduced boundary artifacts. SIAM J. Sci. Comput. **38**, B164–B189 (2016)
9. Cai, J.F., Osher, S., Shen, Z.: Linearized Bregman iterations for frame-based image deblurring. SIAM J. Imaging Sci. **2–1**, 226–252 (2009)
10. Chan, R.H., Riemenschneider, S.D., Shen, L., Shen, Z.: Tight frame: an efficient way for high-resolution image reconstruction. Appl. Comput. Harmon. Anal. **17**, 91–115 (2004)
11. Chan, R.H., Ng, M.K.: Conjugate gradient method for toeplitz systems. SIAM Rev. **38**, 427–482 (1996)
12. Daubechies, I., Defrise, M., De Mol, C.: An iterative thresholding algorithm for linear inverse problems with a sparsity constraint. Commun. Pure Appl. Math. **57–11**, 1413–1457 (2004)
13. Daubechies, I., Han, B., Ron, A., Shen, Z.: Framelets: MRA-based constructions of wavelet frames. Appl. Comput. Harmon. Anal. **14**, 1–46 (2003)
14. Dell'Acqua, P.: A note on Taylor boundary conditions for accurate image restoration. Adv. Comput. Math. **43**, 1283–1304 (2017)
15. Dell'Acqua, P., Donatelli, M., Estatico, C., Mazza, M.: Structure preserving preconditioners for image deblurring. J. Sci. Comput. **72**(1), 147–171 (2017)
16. Dell'Acqua, P., Donatelli, M., Estatico, C.: Preconditioners for image restoration by reblurring techniques. J. Comput. Appl. Math. **272**, 313–333 (2014)
17. Donatelli, M.: Fast transforms for high order boundary conditions in deconvolution problems. BIT **50–3**, 559–576 (2010)
18. Donatelli, M., Estatico, C., Martinelli, A., Serra-Capizzano, S.: Improved image deblurring with anti-reflective boundary conditions and re-blurring. Inverse Probl. **22**, 2035–2053 (2006)
19. Donatelli, M., Hanke, M.: Fast nonstationary preconditioned iterative methods for ill-posed problems, with application to image deblurring. Inverse Probl. **29**, 095008 (2013)
20. Donatelli, M., Martin, D., Reichel, L.: Arnoldi methods for image deblurring with anti-reflective boundary conditions. Appl. Math. Comput. **253**, 135–150 (2015)
21. Donatelli, M., Serra-Capizzano, S.: Antireflective boundary conditions for deblurring problems. J. Electr. Comput. Eng. **2010**, Article ID 241467, 18 (survey) (2010)
22. Engl, H.W., Hanke, M., Neubauer, A.: Regularization Methods for Inverse Problems. Kluwer, Dordrecht (1996)

23. Egger, H., Neubauer, A.: Preconditioning Landweber iteration in Hilbert scales. Numer. Math. **101**, 643–662 (2005)
24. Figueiredo, M., Nowak, R.: An EM algorithm for wavelet-based image restoration. IEEE Trans. Image Process. **12–8**, 906–916 (2003)
25. Fan, Y.W., Nagy, J.G.: Synthetic boundary conditions for image deblurring. Linear Algebra Appl. **434**, 2244–2268 (2011)
26. Hanke, M., Hansen, P.C.: Regularization methods for large-scale problems. Surv. Math. Indust. **3**, 253–315 (1993)
27. Hanke, M., Nagy, J.: Restoration of atmospherically blurred images by symmetric indefinite conjugate gradient techniques. Inverse Probl. **12**, 157–173 (1996)
28. Hanke, M., Nagy, J., Plemmons, R.: Preconditioned iterative regularization for ill-posed problems. In: Numerical Linear Algebra. Proceedings of the Conference in Numerical Linear Algebra and Scientific Computation, Kent, Ohio, March 13–14 1992, de Gruyter, pp. 141–163 (1993)
29. Hansen, P.C., Nagy, J., O'Leary, D.P.: Deblurring Images Matrices, Spectra and Filtering. SIAM Publications, Philadelphia (2005)
30. Kilmer, M.E.: Cauchy-like preconditioners for two-dimensional ill-posed problems. SIAM J. Matrix Anal. Appl. **20**, 777–799 (1999)
31. Nagy, J.G., Palmer, K., Perrone, L.: Iterative methods for image deblurring: a MATLAB object oriented approach. Numer. Algorithms **36** 73–93 (2004). See also: http://www.mathcs.emory.edu/~nagy/RestoreTools
32. Piana, M., Bertero, M.: Projected Landweber method and preconditioning. Inverse Probl. **13–2**, 441–464 (1997)
33. Serra-Capizzano, S.: A note on anti-reflective boundary conditions and fast deblurring models. SIAM J. Sci. Comput. **25**(4) 1307–1325 (2003)
34. Wang, Z., Bovik, A.C., Sheikh, H.R., Simoncelli, E.P.: Image quality assessment: from error visibility to structural similarity. IEEE Trans. Image Process. **13**(4), 600–612 (2004)
35. Yin, W., Osher, S., Goldfarb, D., Darbon, J.: Bregman iterative algorithms for ℓ_1-minimization with applications to compressed sensing. SIAM J. Imaging Sci. **1**, 143–168 (2008)

Non-stationary Structure-Preserving Preconditioning for Image Restoration

Pietro Dell'Acqua, Marco Donatelli and Lothar Reichel

Abstract Non-stationary regularizing preconditioners have recently been proposed for the acceleration of classical iterative methods for the solution of linear discrete ill-posed problems. This paper explores how these preconditioners can be combined with the flexible GMRES iterative method. A new structure-respecting strategy to construct a sequence of regularizing preconditioners is proposed. We show that flexible GMRES applied with these preconditioners is able to restore images that have been contaminated by strongly non-symmetric blur, while several other iterative methods fail to do this.

Keywords Image deblurring · Non-stationary preconditioning · Flexible GMRES

1 Introduction

We are concerned with the restoration of blurred and noise-corrupted images in two space-dimensions. The blurring is modeled by a convolution and the image degradation model is of the form

$$g(\boldsymbol{x}) = [Kf](\boldsymbol{x}) + \nu(\boldsymbol{x}) = \int_{\mathbb{R}^2} h(\boldsymbol{x} - \boldsymbol{y}) f(\boldsymbol{y}) \mathrm{d}\boldsymbol{y} + \nu(\boldsymbol{x}), \qquad \boldsymbol{x} \in \Omega \subset \mathbb{R}^2, \quad (1)$$

P. Dell'Acqua (✉)
Facoltà di Scienze e Tecnologie Informatiche, Libera Università di Bolzano, Bolzano, Italy
e-mail: pietro.dellacqua@gmail.com

M. Donatelli
Dipartimento di Scienza e Alta Tecnologia, Università degli Studi dell'Insubria, Como, Italy
e-mail: marco.donatelli@uninsubria.it

L. Reichel
Department of Mathematical Sciences, Kent State University, Kent, OH, USA
e-mail: reichel@math.kent.edu

© Springer Nature Switzerland AG 2019
M. Donatelli and S. Serra-Capizzano (eds.), *Computational Methods for Inverse Problems in Imaging*, Springer INdAM Series 36,
https://doi.org/10.1007/978-3-030-32882-5_3

where f represents the (desired but unavailable) exact image, h the space invariant point-spread function (PSF) with compact support, ν random noise, and g the (available) blurred and noise-corrupted image. Hence, f and g are real-valued nonnegative functions that determine the light intensity of the desired and available images, respectively.

Discretization of the integral equation (1) at equidistant nodes gives the linear system of algebraic equations

$$g_i = \sum_{j \in \mathbb{Z}^2} h_{i-j} f_j + \nu_i, \qquad i \in \mathbb{Z}^2. \tag{2}$$

The entries of the discrete images $g = [g_i]$ and $f = [f_j]$ represent the light intensity at each picture element (pixel) and $\nu = [\nu_i]$ models the noise-contamination at these pixels. The pixels with index $i \in [1, n]^2$ make up the finite field of view (FOV), which for notational simplicity is assumed to be square. We would like to determine an accurate approximation of the exact image f in the FOV given $h = [h_i]$, distributional information about ν, and the blurred image g in the FOV.

The linear system of algebraic equations defined by (2) with i restricted to $[1, n]^2$ is underdetermined when there are non-vanishing coefficients h_i with $i \neq 0$, because then there are n^2 equations, while the number of unknowns is larger. A common approach to determine a meaningful solution of this kind of underdetermined system is to impose boundary conditions on the image to obtain a linear system of algebraic equations with a square matrix,

$$A f = g, \qquad A \in \mathbb{R}^{n^2 \times n^2}, \qquad f, g \in \mathbb{R}^{n^2}. \tag{3}$$

The boundary conditions specify that the f_j-values in (2) at pixels outside the FOV are linear combinations of f_j-values at certain pixels inside the FOV. Popular boundary conditions include zero Dirichlet boundary conditions (ZDBCs), periodic boundary conditions (PBCs), reflective boundary conditions (RBCs) discussed in [28], and anti-reflective boundary conditions (ARBCs) proposed in [33]. Detailed descriptions and analyses of these boundary conditions can be found in [15, 24, 25]. This paper focuses on ARBCs, which yield an accurate model and often allow simple implementation. We restrict our attention to ARBCs only for the sake of simplicity, but we remark that more accurate boundary conditions and other strategies for dealing with boundary artifacts recently have been proposed in the literature, see [6, 10, 18, 29], and can be applied to construct preconditioners as well. Theoretical results on optimal preconditioning for ARBCs are discussed in [9].

Due to the space-invariance of the PSF, the matrix A has a block Toeplitz-type structure. The detailed structure depends on the boundary conditions. For instance, ZDBCs give a block-Toeplitz–Toeplitz-block (BTTB) structure, while PBCs make A a block-circulant-circulant-block (BCCB) matrix. This is discussed in more detail below.

Quadrantally symmetric PSFs, i.e., PSFs that are symmetric with respect to both the horizontal and vertical axes, arise, e.g., when modeling symmetric Gaussian blur. The associated matrix A in (3) allows diagonalization by a fast transform when RBCs or ARBCs are imposed. These transforms can be applied to develop fast methods for the approximate solution of (3); see [1, 5, 28].

For symmetric PSFs, the matrix A is symmetric and many iterative regularization methods can be applied to the approximate solution of (3), such as non-stationary iterative methods and variants of the minimal residual method; see [12, 16]. On the other hand, for strongly non-symmetric PSFs specially designed iterative regularization methods have to be applied; see [14, 20] for illustrations.

The matrix A in (3) generally is severely ill-conditioned and may be numerically rank-deficient. We refer to linear system of equations (3) with such a matrix as linear discrete ill-posed problems. Due to the error in the right-hand side vector g in (3), which is caused by the noise ν in (2), and because of the ill-conditioning of the matrix, one generally is not interested in the exact solution of (3) (if it exists). Instead one typically would like to compute a suitable approximate solution that furnishes an accurate approximation of the desired image f. Such an approximate solution can be computed by regularizing the system of equations (3), e.g., by replacing this system by a nearby one, whose solution is less sensitive to the error in g. Regularization methods require the choice of a regularization parameter that determines the amount of regularization.

The present paper is concerned with the development of fast and stable iterative regularization methods for the approximate solution of (3) when the matrix A is defined by a non-symmetric PSF h with ARBCs. In particular, we focus on GMRES-type iterative methods. The (standard) GMRES method is commonly used for the iterative solution of large linear systems of equations with a square non-symmetric matrix that is obtained by the discretization of a well-conditioned problem, such as an elliptic partial differential equation with Dirichlet boundary conditions. In this context, preconditioners are employed to accelerate the convergence of the iterative method. An advantage of GMRES, when compared to other iterative methods such as CGLS, is that GMRES does not require the evaluation of matrix-vector products with A^T, the transpose of A. This is commented on further in Sect. 3.

Preconditioners applied to the iterative solution of linear discrete ill-posed problems (3) should avoid propagating the error ν in g into the the computed approximate solution. We will show that such preconditioners can be determined by incorporating a threshold parameter in their definition. Note that the preconditioning strategy for GMRES proposed in [14] can determine accurate restorations, but may require many iterations when the noise level is low. In fact, typically linear discrete ill-posed problems of the form (3) are more difficult to solve when the noise level is low than when it is high, because the restoration of the former kind of images generally requires more iterations.

We would like to investigate the use of non-stationary preconditioning with GMRES-type methods with the aim to obtain accurate restorations within only a few iterations also when the noise level is low and the PSF is strongly non-symmetric. Instead of solving right-preconditioned systems of the form

$$APz = g, \quad z = P^{-1}f \tag{4}$$

by GMRES, we propose to use the flexible GMRES (F-GMRES) method first described by Saad [30] to solve, at step k,

$$AP_k z = g, \quad z = P_k^{-1}f, \tag{5}$$

where the preconditioner P_k is modified in each iteration. The application of F-GMRES to the solution of linear discrete ill-posed problems has previously been discussed by Gazzola and Nagy [19] and Morikuni et al. [27]. The preconditioners developed in the present paper are new. Exploiting the tools developed within the framework of preconditioned Landweber iterative methods [7, 8], we define preconditioners for F-GMRES. By using a suitable sequence of preconditioners P_k in (5), we obtain a preconditioned F-GMRES method that is well suited for image restoration. The preconditioner P_k depends on a thresholding parameter α_k, whose choice will be discussed in Sect. 3.

Several other preconditioning techniques for linear systems of algebraic equations that arise in image restoration and have a square BTTB-type matrix have been described in the literature; see, e.g., [7, 14, 17, 20, 22, 23] and references therein. In all available preconditioning techniques, the preconditioner P is chosen before the iterations are begun and kept fixed during the computations. This corresponds to applying an iterative method to the preconditioned system (4). A nice recent survey is presented by Gazzola et al. [21]. However, it may be difficult to choose a suitable preconditioner before the start of the iterations. Our approach circumvents this complication by allowing the preconditioner to be updated during the solution process.

This paper is organized as follows. Section 2 contains a brief overview of anti-reflective boundary conditions. A discussion on iterative methods, the construction of our preconditioner, and an introduction of the preconditioned F-GMRES method can be found in Sect. 3. Numerical results are presented in Sects. 4, and 5 contains concluding remarks.

2 Anti-reflective Boundary Conditions

We review some properties of blurring matrices with ARBCs. A survey of blurring matrices with ARBCs is given in [15], where many details are provided. Consider a blurring matrix A determined by a discretized PSF,

$$H = \begin{bmatrix} h_{-m,-m} & \cdots & & h_{-m,0} & & \cdots & h_{-m,m} \\ & & & & & & \\ \vdots & \ddots & & \vdots & & & \vdots \\ & & h_{-1,-1} & h_{-1,0} & h_{-1,1} & & \\ h_{0,-m} & \cdots & h_{0,-1} & h_{0,0} & h_{0,1} & \cdots & h_{0,m} \\ & & h_{1,-1} & h_{1,0} & h_{1,1} & & \\ \vdots & & & \vdots & & \ddots & \vdots \\ h_{m,-m} & \cdots & & h_{m,0} & & \cdots & h_{m,m} \end{bmatrix} \in \mathbb{R}^{(2m+1)\times(2m+1)}, \quad (6)$$

with $h_{0,0}$ the central coefficient and, generally, $2m + 1 \ll n$. The image values $f_{1-j,t}$ for $1 \le j \le m$ and $1 \le t \le n$ are represented by $2f_{1,t} - f_{j+1,t}$. Similarly, for $1 \le j \le m$ and $1 \le s, t \le n$, we obtain the image values

$$f_{s,1-j} = 2f_{s,1} - f_{s,j+1}, \quad f_{n+j,t} = 2f_{n,t} - f_{n-j,t}, \quad f_{s,n+j} = 2f_{s,n} - f_{s,n-j}.$$

When both indices of $f_{p,q}$ are outside the range $\{1, 2, \ldots, n\}$, which happens for pixels close to the four corners of the given image, we carry out anti-reflection first in one space-direction (in the direction of the horizontal or vertical axis) and then in the other direction; see [13]. We describe these anti-reflections for pixels near the corner with pixel index $(1, 1)$ of an image; pixels near the other corners are treated analogously. Thus, for $1 \le j, l \le m$, we let

$$f_{1-j,1-l} = 4f_{1,1} - 2f_{1,l+1} - 2f_{j+1,1} + f_{j+1,l+1}.$$

Here we have carried out anti-reflection along the horizontal axis followed by anti-reflection along the vertical axis.

The strategy to anti-reflect first in one space-direction and then in an orthogonal space-direction yields a blurring matrix $A \in \mathbb{R}^{n^2 \times n^2}$ with a two-level structure; see [13]. Specifically, A is the sum of five matrices: A block Toeplitz matrix with Toeplitz blocks, a block Toeplitz matrix with Hankel blocks, a block Hankel matrix with Toeplitz blocks, a block Hankel matrix with Hankel blocks, and a matrix of rank at most $4n$. Despite this somewhat complicated structure, matrix-vector products with the matrix A can be evaluated in $\mathcal{O}(n^2 \log(n))$ arithmetic floating-point operations (flops) by applying the FFT as follows: Let the n^2-vectors $x = X(:)$ and $y = Y(:)$ be defined by stacking the columns of the $n \times n$-matrices X and Y, respectively. These matrices represent images; their entries are pixel values. For every kind of boundary conditions, the matrix-vector product $y = Ax$ can be implemented by the following procedure:

1. pad X with the chosen boundary conditions to obtain an extended 2D array $\tilde{X} \in \mathbb{R}^{(n+m)\times(n+m)}$;
2. compute $\tilde{Y}$ as the circular convolution of $\tilde{X}$ and H;
3. determine Y by extracting the central inner $n \times n$ part of $\tilde{Y}$.

The anti-reflective boundary conditions require the use of an anti-symmetric pad analogous to the symmetric pad that is available for the MATLAB function padarray.[1] Further details are provided in [15].

3 The Preconditioned Iterative Method

This section defines the preconditioners to be used and discusses the iterative solution of the preconditioned linear systems of algebraic equations (5) by the F-GMRES method.

3.1 Iterative Regularization Methods for Anti-reflective Boundary Conditions

Introduce the correlation operator

$$[K^*f](x) = \int_{\mathbb{R}^2} h(y - x) f(y) \mathrm{d}y, \tag{7}$$

which is the adjoint of the convolution operator in (1). Here we have used the fact that h is real-valued. Let the matrix A' be obtained by discretizing (7) with the same boundary conditions as for (3). It is proposed in [11] that, instead of solving (3), one should compute the solution of the linear system of equations

$$A'Af = A'g \tag{8}$$

when RBCs or ARBCs are imposed and the PSF is quite general, such as a PSF that models motion blur. The linear system (8) is solved by a conjugate gradient (CG) method that is formally similar to the CGLS method [2]. The latter method computes an approximate solution of (3) by determining an approximate solution of the associated normal equations $A^TAf = A^Tg$. It is suggested in [11] that the matrix A^T in the CGLS method be replaced by A'. Attractions of the so obtained iterative method for the approximate solution of (3) include that the matrix $A'A$ is not explicitly formed (only matrix vector products with the matrices A and A' are evaluated) and that the method uses short recurrence relations. Therefore, the method requires fairly little computer storage. The reason for using A' instead of A^T in (8) is that the evaluation of matrix-vector products with the latter matrix is more cumbersome and may suffer from numerical instability; see, e.g., [14] for a recent discussion.

---

[1]A MATLAB code for the anti-symmetric pad can be downloaded at http://scienze-como. uninsubria.it/mdonatelli/Software/software.html.

However, the iterative solution of (8) by the CG method is not without difficulties. The main problem is that the matrix $A'A$ is not symmetric positive definite and, therefore, the application of the CG method to the solution of (8) has no theoretical justification. Moreover, the computed restorations may be of poor quality, in particular when the PSF is strongly non-symmetric; see the analysis in [14]. These difficulties with the CG method prompted the investigation in [14] of the application of GMRES-type methods to the solution of (8).

GMRES is an iterative method proposed in [32] for the solution of linear systems of algebraic equations with a fairly general square non-symmetric non-singular matrix; see also [31]. Among several solution methods investigated in [14], the application of GMRES to the system

$$AA'z = g \tag{9}$$

performed the best. When z is an approximate solution of (9), $f = A'z$ is an approximate solution of (3). We may consider A' a right preconditioner. Right-preconditioning is convenient to use when the number of iterations is determined with the aid of the discrepancy principle; see below. We next describe several preconditioners.

To justify the definition of our preconditioner, we first discuss left-preconditioned Landweber iteration. Following [7], where the so called "Z variant" is described, we consider the left-preconditioned system

$$ZAf = Zg \tag{10}$$

obtained from (3). Application of Landweber iteration to the solution of (10) yields the iterates

$$f_{k+1} = f_k + Z(g - Af_k). \tag{11}$$

We may determine the preconditioner $Z \in \mathbb{R}^{n^2 \times n^2}$ by filtering as follows: In the case of PBCs, A is a BCCB matrix, which can be diagonalized by the 2D discrete Fourier transform. The eigenvalues $\lambda_{i,j}$ of A, for $i, j = 0, \ldots, n - 1$, can be computed by the 2D FFT applied to its first column arranged as a 2D array. The matrix Z is chosen to be a BCCB matrix, whose eigenvalues $\check{\lambda}_{i,j}$ are obtained by applying some filter to the $\lambda_{i,j}$. For instance, we may use a slightly modified version of the Tikhonov filter

$$\check{\lambda}_{i,j} = \frac{\overline{\lambda}_{i,j}}{\left|\lambda_{i,j}\right|^2 + \alpha}, \qquad i, j = 0, 1, \ldots, n - 1,$$

where $\alpha > 0$ is a regularization parameter and the bar denotes complex conjugation. The BCCB matrix Z can be defined analogously for other boundary conditions; see [7] for details.

When applying a stationary preconditioned iterative regularization method, we have to face the non-trivial task of determining a suitable value of the parameter α. A too small value of α often gives fast convergence, but may cause instability due to

severe ill-conditioning of the preconditioner. The instability may result in large prop-
agated errors, which may reduce the quality of the computed solution and possibly
make the computed solution useless. On the other hand, a too large value of α may
result in slow convergence of the computed iterates. Hence, a proper choice of α is
important. Since iterative regularization methods are filtering methods, and the filter
changes with the iteration number, it can be difficult to determine a priori a value of
α that is suitable for all iterations. Here we also note that the number of iterations
required is typically not known before the iterative solution process is started.

To avoid the task of determining a suitable value of α before the start of the
iterations, Donatelli and Hanke [12] proposed the following non-stationary version
of the iterations (11),

$$\boldsymbol{f}_{k+1} = \boldsymbol{f}_k + Z_{\text{circ}}^k \boldsymbol{r}_k, \quad Z_{\text{circ}}^k = C^T (CC^T + \alpha_k I)^{-1}, \quad \boldsymbol{r}_k = \boldsymbol{g} - A\boldsymbol{f}_k. \quad (12)$$

Here C is the BCCB matrix associated with the PSF that defines the matrix A in (3)
and α_k is determined by solving a non-linear equation by Newton's method; see [12]
for details.

Recently, Dell'Acqua et al. [8] extended the non-stationary method (12) to be
able to take the boundary conditions of the problem into account and proposed the
following iteration scheme,

$$\boldsymbol{f}_{k+1} = \boldsymbol{f}_k + Z_{\text{struct}}^k \boldsymbol{r}_k, \quad Z_{\text{struct}}^k = \mathcal{B}(C^T (CC^T + \alpha_k I)^{-1}), \quad \boldsymbol{r}_k = \boldsymbol{g} - A\boldsymbol{f}_k, \quad (13)$$

where the operator $\mathcal{B}$ denotes the application of boundary conditions to the circulant
matrix $C^T (CC^T + \alpha_k I)^{-1}$. Thus, the operator $\mathcal{B}$ affects the structure of the matrix
Z_{struct}^k.

The matrix Z_{struct}^k may be considered a preconditioner. In particular, we may solve
the right-preconditioned linear system (5) with the preconditioner

$$P_k = \mathcal{B}(C^T (CC^T + \alpha_k I)^{-1}) \quad (14)$$

by F-GMRES. The parameter α_k allows the preconditioner P_k to be varied during
the iterations.

Note that the preconditioner (14) is not explicitly formed, only matrix-vector
products with P_k are computed. Indeed, the matrix C is not explicitly formed; instead
matrix-vector products are evaluated by circular convolutions with the coefficient
mask H, i.e., with the PSF. The same can be done for Z_{circ}^k and Z_{struct}^k. We remark that
the structure of Z_{struct}^k may be quite involved depending on the boundary conditions,
but, using the procedure described at the end of Sect. 2, only a 2D coefficient mask
$\check{H}$ is required to be used instead of H. The procedure for computing $\check{H}$ is described
in the next subsection.

3.2   Construction of the Preconditioners

The generating function associated with the PSF defined by H in (6) is given by

$$f(x_1, x_2) = \sum_{j_1=-m}^{m} \sum_{j_2=-m}^{m} h_{j_1, j_2} e^{\hat{\imath}(j_1 x_1 + j_2 x_2)}, \qquad \hat{\imath} = \sqrt{-1}. \tag{15}$$

Thus, the entries h_{j_1, j_2} of the matrix H are Fourier coefficients of the function f.

Let $y_k^{(n)} = 2\pi k/n$, for $k = 0, \ldots, n-1$, be a uniform sampling on the interval $[0, 2\pi]$. The 2D Fourier matrix of order $n^2 \times n^2$ is given by

$$F_{(n,n)} = F_n \otimes F_n, \quad \text{where} \quad F_n = \frac{1}{\sqrt{n}} \left[e^{-\hat{\imath} j y_k^{(n)}} \right]_{k,j=0}^{n-1}$$

and $\otimes$ denotes the Kronecker product. Given the function f in (15), the BCCB matrix C generated by f is defined as

$$C = \mathcal{C}_{(n,n)}(f) = F_{(n,n)} D_{(n,n)}(\lambda) F_{(n,n)}^H,$$

where $D_{(n,n)}(\lambda) = \text{diag}_{i,j=0,\ldots,n-1}[\lambda_{i,j}]$ is the diagonal matrix of its eigenvalues and $F_{(n,n)}^H$ is the conjugate transpose of $F_{(n,n)}$. The eigenvalues $\lambda_{i,j}$, $0 \le i, j < n$, of C are determined by a uniform sampling of the generating function f in (15) at the grid points $\Gamma_n = \{(y_i^{(n)}, y_j^{(n)}) : i, j = 0, 1, \ldots, n-1\}$, namely

$$\lambda_{i,j} = f\left(\frac{2\pi i}{n}, \frac{2\pi j}{n}\right), \qquad i, j = 0, 1, \ldots, n-1. \tag{16}$$

Therefore, the PSF can be interpreted as a mask of Fourier coefficients, and the BCCB matrix C generated by f in (15) coincides with the matrix A in (3) when PBCs are imposed.

The preconditioner is constructed by using the Tikhonov filter, but other filters can be applied as well. The Tikhonov solution of the linear system $Cf = g$ is

$$f_\alpha = (C^T C + \alpha I)^{-1} C^T g = C^T (CC^T + \alpha I)^{-1} g.$$

The BCCB matrix $C^T(CC^T + \alpha I)^{-1}$ has the eigenvalues

$$\check{\lambda}_{i,j} = \frac{\bar{\lambda}_{i,j}}{|\lambda_{i,j}|^2 + \alpha} = \frac{\overline{f}(\frac{2\pi i}{n}, \frac{2\pi j}{n})}{|f(\frac{2\pi i}{n}, \frac{2\pi j}{n})|^2 + \alpha}, \qquad i, j = 0, 1, \ldots, n-1. \tag{17}$$

Similarly to (16), assuming for simplicity that n is odd, the eigenvalues $\check{\lambda}_{i,j}$ may be considered a sampling of the function

$$g(x_1, x_2) = \sum_{j_1, j_2 = -\frac{n-1}{2}}^{\frac{n-1}{2}} \beta_{j_1, j_2} e^{i(j_1 x_1 + j_2 x_2)}$$

at the grid points Γ_n for the specific choice of the coefficients β_{j_1, j_2} as now discussed. The trigonometric polynomial g is determined by the n^2 interpolation conditions $\check{\lambda}_{i,j} := g\left(\frac{2\pi i}{n}, \frac{2\pi j}{n}\right)$ and its coefficients β_{j_1, j_2} can be computed by means of a two-dimensional IFFT. Note that g is a regularized approximation of the inverse of f on Γ_n. Let $\check{H}$ denote the mask for the Fourier coefficients β_{j_1, j_2}. It can be determined by carrying out the following steps:

1. Compute $\lambda_{i,j}$ in (16) by the FFT applied to H.
2. Compute $\check{\lambda}_{i,j}$ in (17).
3. Compute $\check{H}$ by the IFFT applied to $\check{\lambda}_{i,j}$.

In actual computations, we modify the BCCB matrix $C^T(CC^T + \alpha I)^{-1}$ to correspond to ARBCs. This yields a structured preconditioner $P = \mathcal{B}(C^T(CC^T + \alpha I)^{-1})$, where $\mathcal{B}$ is an operator that imposes the ARBCs. As already mentioned at the end of Sect. 2, the matrix P is not explicitly formed, but only $\check{H}$ is stored and a matrix-vector product with P is evaluated in $\mathcal{O}(n^2 \log(n))$ flops by using the anti-symmetric pad and convolution with $\check{H}$.

3.3   The Flexible GMRES Method

The F-GMRES method [30] is a minimal residual iterative method that is designed for application of a sequence of preconditioners. We will use F-GMRES with preconditioners of the form (14) that are determined by a sequence of α_k-values.

Given a set of ℓ linearly independent vectors $u_1, u_2, \ldots, u_\ell \in \mathbb{R}^{n^2}$, the F-GMRES method determines a decomposition of the form

$$AU_\ell = V_{\ell+1} H_{\ell+1, \ell}, \tag{18}$$

where $U_\ell = [u_1, u_2, \ldots, u_\ell] \in \mathbb{R}^{n^2 \times \ell}$, $V_{\ell+1} = [v_1, v_2, \ldots, v_{\ell+1}] \in \mathbb{R}^{n^2 \times (\ell+1)}$ has orthonormal columns with $v_1 = g/\|g\|$, and $H_{\ell+1, \ell} = [h_{i,j}] \in \mathbb{R}^{(\ell+1) \times \ell}$ is of upper Hessenberg type. Let $e_1 = [1, 0, \ldots, 0]^T \in \mathbb{R}^{\ell+1}$ denote the first axis vector. Then

$$\min_{w \in \text{range}(U_\ell)} \|Aw - g\| = \min_{y \in \mathbb{R}^\ell} \|AU_\ell y - g\| = \min_{y \in \mathbb{R}^\ell} \|H_{\ell+1, \ell} y - e_1 \|g\| \|, \tag{19}$$

where $\| \cdot \|$ denotes the Euclidean vector norm.

Assume that the matrix $H_{\ell+1, \ell}$ exists and that all its subdiagonal entries are positive. This is the generic situation. The positivity of the subdiagonal entries of the upper Hessenberg matrix $H_{\ell+1, \ell}$ secures that its columns are linearly independent. We remark that the parameter ℓ in (18) generally is fairly small in our applications.

The minimization problem on the right-hand side of (19) has a unique solution $y_\ell \in \mathbb{R}^\ell$, which determines the approximate solution $f_\ell = U_\ell y_\ell$ of (3). Since ℓ is small, the solution y_ℓ easily can be computed by QR factorization of the matrix $H_{\ell+1,\ell}$. In our application of the decomposition (18), the vectors u_k are determined with the preconditioners P_k; see (14) and Algorithm 1 below. A discussion on the choice of the parameters α_k in the preconditioner P_k and on the choice of ℓ in (18) is provided in Sect. 3.4.

Algorithm 1 *The F-GMRES method*
1. $v_1 = g/\|g\|$
2. for $k = 1, 2, \ldots, \ell$ do
3. $u_k = P_k v_k; \; v = A u_k$
4.   for $i = 1, 2, \ldots, k$ do
5. $h_{i,k} = v^T v_i; \; v = v - h_{i,k} v_i$
6.   end
7. $h_{k+1,k} = \|v\|; \; v_{k+1} = v/h_{k+1,k}$
8. end
9. define $U_\ell = [u_1, u_2, \ldots, u_\ell] \in \mathbb{R}^{n^2 \times \ell}$ and $H_{\ell+1,\ell} = [h_{i,j}] \in \mathbb{R}^{(\ell+1) \times \ell}$ upper Hessenberg
10. compute $y_\ell := \arg\min_{y \in \mathbb{R}^\ell} \|H_{\ell+1,\ell} y - e_1 \|g\| \|$ and $f_\ell = U_\ell y_\ell$

We say that F-GMRES breaks down at step k if $h_{k+1,k} = 0$ and $h_{j+1,j} > 0$ for $1 \le j < k$. As already mentioned, this is a rare event. Discussions on breakdown of F-GMRES can be found in [27, 30].

Algorithm 1 requires that both the matrix $V_{\ell+1}$ and the vectors $u_k = P_k v_k$, $1 \le k \le \ell$, be stored. This implies that F-GMRES demands more storage space than (standard) GMRES after the same number of steps, since GMRES only requires storage of the matrix $V_{\ell+1}$. However, the fact that F-GMRES allows non-stationary preconditioning, while GMRES does not, may be worth the extra storage cost. Note that if we let $P_k = P$ be independent of k, then Algorithm 1 can be replaced by the preconditioned (standard) GMRES method; see [31] for a discussion of the latter.

A difference between the F-GMRES and GMRES algorithms is that the action of $A P_k$ on a vector v generally is not in the range of $V_{\ell+1}$ in F-GMRES. Instead, we have the following result, which is a consequence of (19).

Proposition 1 *The approximate solution f_ℓ obtained at step ℓ of F-GMRES minimizes the residual norm $\|g - A f_\ell\|$ over* range(U_ℓ).

Another difference is that, while for standard GMRES with initial iterate in span$\{g\}$ breakdown is equivalent to convergence, this is not the case for F-GMRES. Moreover, it is difficult to show convergence results for F-GMRES since, differently from standard GMRES, there is no isomorphism between the solution subspace of F-GMRES and the space of polynomials. For more information, we refer to [27, 30, 31].

3.4 The Stopping Criterion and the Choice of Regularization Parameters

This subsection discusses how to determine the number of iterations, ℓ, with Algorithm 1 and how to choose the parameters α_k of the preconditioners P_k; see (14). We will assume that a fairly accurate bound ε for the norm of the error $\boldsymbol{\nu}$ in the vector $\boldsymbol{g}$ is available. Thus,

$$\|\boldsymbol{\nu}\| \leq \varepsilon.$$

Let $\boldsymbol{f}_1, \boldsymbol{f}_2, \boldsymbol{f}_3, \ldots$ be a sequence of approximate solutions of (3) determined by Algorithm 1 and define the associated residual vectors $\boldsymbol{r}_\ell = \boldsymbol{g} - A\boldsymbol{f}_\ell, \ell = 1, 2, 3, \ldots$. The discrepancy principle prescribes that the iterations with Algorithm 1 be terminated as soon as a residual vector $\boldsymbol{r}_\ell$ that satisfies

$$\|\boldsymbol{r}_\ell\| \leq \eta\varepsilon \tag{20}$$

has been determined, where $\eta \geq 1$ is a user-specified constant independent of ε. We set $\eta = 1$ in the computed examples reported in Sect. 4. This stopping criterion is reasonable, because the desired exact solution $\boldsymbol{f}$ satisfies $\|A\boldsymbol{f} - \boldsymbol{g}\| = \|\boldsymbol{\nu}\|$. Note that since

$$\|\boldsymbol{r}_\ell\| = \|H_{\ell+1,\ell}\boldsymbol{y}_\ell - \boldsymbol{e}_1\|\boldsymbol{g}\|\,\|,$$

we can check whether (20) holds without explicitly forming the residual vector $\boldsymbol{r}_\ell$.

We use a progressive implementation of F-GMRES in the computed examples. This implementation updates the matrix $H_{\ell+1,\ell}$ for $\ell = 1, 2, 3, \ldots$ together with its QR factorization. The solution $\boldsymbol{y}_\ell$ of the small least-squares problem in the right-hand side of (19) is computed for every ℓ. This makes it easy to determine when (20) holds for the first time and, therefore, when the iterations should be terminated.

We turn to the determination of the parameters α_k of the preconditioners P_k, $k = 1, 2, 3, \ldots$. A well-established choice of the regularization parameter in the context of iterated Tikhonov methods is the geometric sequence

$$\alpha_k = \alpha_0 q^{k-1}, \qquad k = 1, 2, 3, \ldots, \tag{21}$$

where $\alpha_0 > 0$ and $0 < q < 1$; see, e.g., [3, 12]. This choice also is used for Bregman iteration [4, 26]. The value of the initial regularization parameter α_0 is not critical as long as it is not too small.

A different technique to determine the regularization parameter α_k is described in [12]: At step k the parameter α_k is determined by solving the non-linear equation

$$\|\boldsymbol{r}_k - CZ_{\text{circ}}^k\boldsymbol{r}_k\| = q_k\|\boldsymbol{r}_k\| \tag{22}$$

with a few steps of Newton's method, where

$$q_k = \max\left\{q, \; 2\rho_{\text{circ}} + (1 + \rho_{\text{circ}})\delta/\|\boldsymbol{r}_k\|\right\} \tag{23}$$

and Z_{circ}^k is defined as in (12). The parameter q is included in (23) as a safeguard to prevent that the q_k decrease too rapidly with increasing k. We remark that the theoretical results in [12] do not use the parameters q in (23). The parameter $0 < \rho_{\text{circ}} < 1/2$ should be as small as possible and satisfy

$$\|(C - A)z\| \leq \rho_{\text{circ}}\|Az\|, \qquad \forall z \in \mathbb{R}^n. \tag{24}$$

If A is accurately approximated by its BCCB counterpart C, then this inequality can be approximately satisfied for a small value of ρ_{circ}. This parameter has to be set in the algorithm described in [12]; for image deblurring problems, it is usually chosen as 10^{-2} or 10^{-3}. A too small value of ρ_{circ} can easily be recognized by an oscillatory behavior of the α_k with increasing values of k; see [12].

In [8], the same approach is applied to the structured case. The goal is to estimate α_k by solving

$$\|\boldsymbol{r}_k - AZ_{\text{struct}}^k \boldsymbol{r}_k\| = q_k\|\boldsymbol{r}_k\| \tag{25}$$

for α_k defining Z_{struct}^k as described by (13). This is not computationally practicable when the PSF has a non-symmetric structure. Instead, the regularization parameter α_k is estimated by using Eq. (22), which may be considered a computable approximation of (25). Note that even though we again use Eq. (22) to estimate the parameters α_k, we obtain a different parameter sequence $\alpha_1, \alpha_2, \alpha_3, \dots$, because the sequence of residual vectors differs. Furthermore, in this case condition (24) is not meaningful. Therefore, a new parameter ρ_{struct} is introduced and the iterations are terminated by the discrepancy principle (20) with

$$\eta = \frac{1 + 2\rho_{\text{struct}}}{1 - 2\rho_{\text{struct}}}. \tag{26}$$

In the following, we refer to the sequence $\alpha_1, \alpha_2, \alpha_3, \dots$ computed in the way described as the DH sequence, since it is a development of an idea introduced in [12]. We set $q = 0.8$ and $\rho_{\text{struct}} = 10^{-2}$ in all the computed examples.

We now describe a new technique for determining the parameters α_k. Assume that

$$\|\boldsymbol{r}_k\| = c_k \alpha_k^p,$$

where $c_k > 0$ and α_k are positive scalars, and $p \geq 1$ is a fixed exponent. Requiring that $\|\boldsymbol{r}_k\| = \varepsilon$ yields

$$\alpha_k = \sqrt[p]{\frac{\varepsilon}{c_k}}.$$

Assuming that

$$c_k \approx c_{k-1} = \frac{\|\boldsymbol{r}_{k-1}\|}{\alpha_{k-1}^p},$$

we obtain the sequence

$$\alpha_1 = \alpha_0,$$

$$\alpha_k = \sqrt[p]{\frac{\varepsilon}{\|r_{k-1}\|}}\, \alpha_{k-1}, \qquad k = 2, 3, \dots, \tag{27}$$

where $\alpha_0 > 0$. In other words, the update from α_{k-1} to α_k is based on the ratio of the error bound ε and the norm of the residual relative to the previous iteration. The heuristic motivation behind this strategy is that α_k is computed in order to make the norm of the residual $\|r_k\|$ move towards ε as k increases. When using (27), the key parameter for the generation of the sequence $\{\alpha_k\}$ is p. In the computed examples, we set $\alpha_0 = 1$ and we illustrate that $p = 2$ is a good choice for all the considered examples.

4   Numerical Results

This section reports some numerical results that show the performance of the methods described. We consider two deblurring problems with non-symmetric PSFs and low noise levels, and apply ARBCs. It is the purpose of the examples to illustrate the efficacy of non-stationary structure-preserving preconditioners for F-GMRES and Landweber iteration. The use of different techniques for computing the sequence of regularization parameters is illustrated.

4.1   Test Problems

Figure 1 displays images for our first test problem (Test 1). The left panel shows the uncontaminated (exact) camera man image. We take a larger image and crop it (see the white box in Fig. 1a). The image inside the box is made up of 227×227 pixels and is assumed not to be available. The PSF, which is three quarters of a Gaussian blur, is shown in the middle panel of Fig. 1. It is made up of 29×29 pixels. The (available) blur- and noise-contaminated image is displayed by the right panel of Fig. 1. It is computed by applying the PSF to the larger image of the left panel, adding 0.5% white Gaussian noise, and then using a 227×227 pixel subimage.

Let the vector $f \in \mathbb{R}^{227^2}$ represent the desired uncontaminated image, where we order the pixels of the image column-wise. Similarly, we let the vector $f_{\text{restored}} \in \mathbb{R}^{227^2}$ represent the restored image determined by one of the methods in our comparison. We refer to

$$\frac{\|f_{\text{restored}} - f\|}{\|f\|}$$

as the relative restoration error (RRE).

(a)   **(b)**   **(c)**

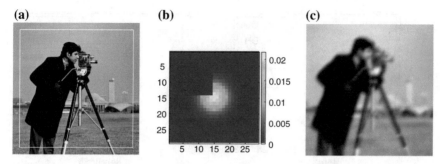

Fig. 1 [Test 1] **a** True image, **b** PSF, **c** blurred and noisy image

(a)   **(b)**   **(c)**

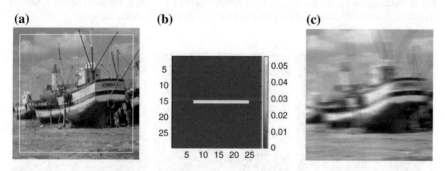

Fig. 2 [Test 2] **a** True image, **b** PSF, **c** blurred and noisy image

Figure 2 shows images for our second test problem (Test 2), and is analogous to Fig. 1. The uncontaminated unavailable (exact) boat image is made up of 227×227 pixels and shown inside the white box of Fig. 2a. The middle panel depicts the non-symmetric PSF. It models motion blur and is made up of 29×29 pixels. The right panel displays the available blur- and noise-contaminated image. The noise is 0.6% white Gaussian.

4.2 Restorations, Plots, and Tables

We compare the unpreconditioned CGLS and GMRES methods to non-stationary preconditioned F-GMRES. Throughout this section CGLS refers to the conjugate gradient method applied to the solution of the linear system of equations (8); see the discussion in Sect. 3.1. Also preconditioned Landweber as described in [8] is considered. The preconditioners are determined by the regularization parameters α_k defined by (21), (22), or (27). We refer to these sequences of regularization parameters as the "geometric sequence", the "DH sequence", and the "new sequence", respectively. We always set $\alpha_0 = 1$ and let $q = 0.8$ for the geometric sequence, $q = 0.8$ and $\rho_{\text{struct}} = 10^{-2}$ for the DH sequence, and $p = 2$ for the new sequence.

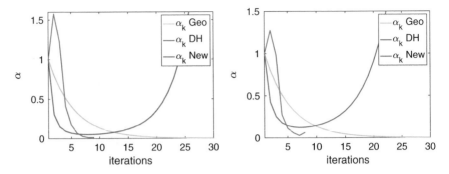

Fig. 3 Plots of the α_k as a function of k related to F-GMRES for the geometric sequence, the DH sequence, and the new sequence for Test 1 (on the left) and Test 2 (on the right)

(a)   **(b)**   **(c)**

Fig. 4 [Test 1] Restorations determined with the discrepancy principle by **a** CGLS (RRE 0.0923, IT 27); **b** F-GMRES New (RRE 0.0907, IT 8); **c** Z^k_{struct}-Landweber New (RRE 0.0942, IT 12)

These three approaches to choosing the α_k are used for both the Z^k_{struct}-Landweber and F-GMRES iterations. Figure 3 shows the α_k for F-GMRES. It can be seen that these three approaches give quite different parameter sequences $\alpha_1, \alpha_2, \alpha_3, \ldots$; the geometric sequence converges to zero at a rate that depends on the choice of q; the DH sequence achieves values larger than unity in the first steps, and then the sequence decreases rapidly. Finally, the sequence (27) decreases quickly to a small value (close to 0) and then increases.

Figures 4 and 5 show (for Test 1 and Test 2, respectively) restorations determined by different methods when the iterations are terminated by the discrepancy principle using (20), and when applicable (26), and the α_k are determined as described in [12]. The restored images look essentially the same; also their RRE values are close. The non-stationary preconditioners of this paper give rapid convergence and restorations of high quality. Tables 1 and 2 report (for Test 1 and Test 2, respectively) the RRE and the number of iterations (IT) required to achieve the best restoration (i.e., the restoration with the smallest RRE) and the restoration determined with the discrepancy principle. The symbol—in the tables indicates that the discrepancy principle cannot be satisfied, while the symbol n/a means that no meaningful best restoration

Fig. 5 [Test 2] Restorations determined with the discrepancy principle by **a** CGLS (RRE 0.0948, IT 17); **b** F-GMRES DH (RRE 0.0925, IT 7); **c** Z^k_{struct}-Landweber DH (RRE 0.0917, IT 12)

Table 1 [Test 1] Relative restoration error (RRE) and iteration number (IT) for the best restoration and when using the discrepancy principle for different methods

Method	RRE best res.	IT	RRE discrepancy	IT
CGLS	0.0895	36	0.0923	27
GMRES	0.1470	5	–	–
F-GMRES Geo	0.0897	13	0.0908	12
F-GMRES DH	n/a	n/a	0.0905	8
F-GMRES New	0.0898	9	0.0907	8
Z^k_{struct}-Landweber Geo	0.0889	23	0.0941	20
Z^k_{struct}-Landweber DH	n/a	n/a	0.0954	13
Z^k_{struct}-Landweber New	0.0888	18	0.0942	12

is available. This situation may arise when the method [12] is applied, because the iterations with this method terminate with the discrepancy principle.

Figures 6 and 7 show (for Test 1 and Test 2, respectively) the RRE as a function of the iteration number. Solid curves are used for CGLS, GMRES, and F-GMRES, while dashed curves are used for Z^k_{struct}-Landweber. For the F-GMRES and Z^k_{struct}-Landweber plots, we use colors to show how the parameter values for the non-stationary preconditioners are determined. The iteration associated with the best restoration is marked by the symbol ∘, while the iteration identified by the discrepancy principle is marked by the symbol ×. Note that for the proposed F-GMRES, the discrepancy principle works very well. The maximum number of iterations is set to 100.

Standard GMRES can be seen to perform poorly for both image restoration problems. The best restoration determined by GMRES has a larger error than the best restoration achieved with any of the other methods in our comparison. Moreover, the discrepancy principle fails to terminate the iterations within 100 steps. Discussions on why GMRES may perform poorly for some restoration problems can be found in [14, 20].

Table 2 [Test 2] Relative restoration error (RRE) and iteration number (IT) for the best restoration and for the iterate determined by the discrepancy principle for different methods

Method	RRE best res.	IT	RRE discrepancy	IT
CGLS	0.0939	15	0.0948	17
GMRES	0.1072	84	–	–
F-GMRES Geo	0.0931	8	0.0935	9
F-GMRES DH	n/a	n/a	0.0925	7
F-GMRES New	0.0932	7	0.0932	7
Z_{struct}^k-Landweber Geo	0.0912	17	–	–
Z_{struct}^k-Landweber DH	n/a	n/a	0.0917	12
Z_{struct}^k-Landweber New	0.0912	12	0.0917	13

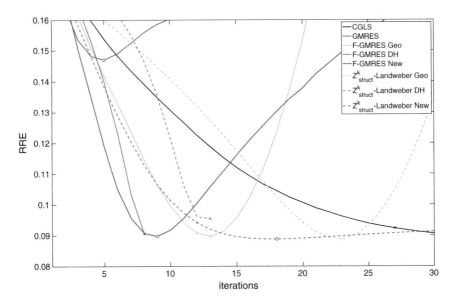

Fig. 6 [Test 1] Plots of the RRE for different methods. The iterate that gives the best restoration is marked by ○, and the iterate determined by the discrepancy principle is marked by ×

The restorations computed by CGLS are quite accurate, in particular for Test 1. However, CGLS requires a large number of iterations in comparison to the other methods considered, where we recall that each iteration with CGLS demands two matrix-vector product evaluations, one with A and one with A'.

Comparing the preconditioned F-GMRES method to the Z_{struct}^k-Landweber method described in [8], we note that the latter usually achieves an insignificantly smaller RRE, but preconditioned F-GMRES requires fewer iterations and terminates reliably with the aid of the discrepancy principle. Note that in Test 2, the discrepancy principle fails to terminate the iterations with Z_{struct}^k-Landweber Geo.

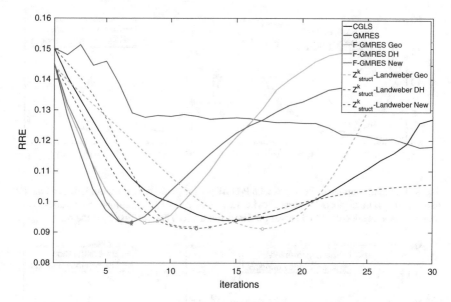

Fig. 7 [Test 2] Plots of RRE for different methods. The iterate that gives the best restoration is marked by ○, and the iterate determined by the discrepancy principle is marked by ×

In summary, the non-stationary preconditioning approach can be used effectively with F-GMRES. Comparing the different ways to determine the parameters α_k for the preconditioners, Figs. 6 and 7 show the sequence (27) to give the fastest convergence in both Test 1 and Test 2 for both the F-GMRES and Z^k_{struct}-Landweber methods.

4.3   Robustness Analysis of P-GMRES

To better justify the non-stationary preconditioning approach of this paper, we present some numerical results for P-GMRES, i.e., GMRES with the same preconditioner for all iterations; all α_k have the same value α in each step. We illustrate that differently from F-GMRES with the different preconditioning strategies previously described, P-GMRES is very sensitive to the choice of α.

Figure 8 displays how α affects the number of iterations required to satisfy the discrepancy principle and to determine the best restoration. We can see that more iterations are required for larger values of α. Moreover, Fig. 8 shows that termination of the iterations with the discrepancy principle does not work well when α is too small. In fact, the number of iterations grows in Test 1 when α is reduced, while in Test 2 the discrepancy principle fails to stop the iterative method for α smaller than 10^{-2}.

The quality of the computed restorations is depicted in Fig. 9. The blue curves show the RRE for the best restorations determined by P-GMRES for different values of α.

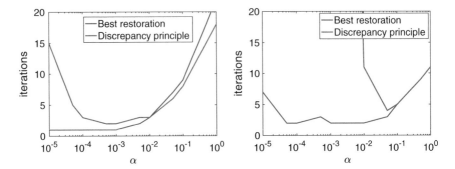

Fig. 8 Plot of the number of iterations with P-GMRES that gives the best restorations as a function of α (blue curve), and plot of the number of iterations with P-GMRES determined by the discrepancy principle as a function of α (red curve) for Test 1 (on the left) and for Test 2 (on the right)

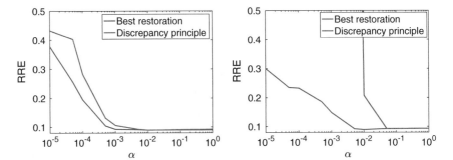

Fig. 9 Plot of the RRE for the best restoration computed by P-GMRES as a function of α (blue curve) and plot of the RRE for the P-GMRES iterate determined by the discrepancy principle as a function of α (red curve) for Test 1 (on the left) and for Test 2 (on the right)

The red curves show the RRE when terminating the iterations with the discrepancy principle. When α is reduced, the quality of restorations quickly deteriorates. We can notice again that the discrepancy principle does not work well; in Test 2 it is not able to stop the method for α smaller than 10^{-2}.

4.4 Robustness Analysis of the DH Sequence

The DH sequence depends on the choice of parameters q and ρ_{struct}. The former parameter is included as a safeguard to prevent the q_k in (22) from decreasing too rapidly. We set $q = 0.8$, the same as for the geometric sequence. This subsection shows the robustness of F-GMRES with respect to the choice of ρ_{struct}.

Figure 10 displays how ρ_{struct} affects the number of iterations required to satisfy the discrepancy principle. As expected, we can see that fewer iterations are required for larger values of ρ_{struct}.

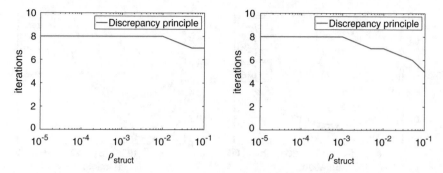

Fig. 10 Plots of the number of iterations by F-GMRES DH determined by the discrepancy principle as a function of ρ_{struct} for Test 1 (on the left) and for Test 2 (on the right)

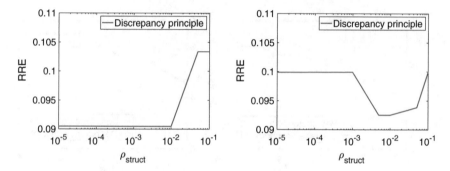

Fig. 11 Plots of the RRE for iterates computed by F-GMRES DH using the discrepancy principle as a function of ρ_{struct} for Test 1 (on the left) and for Test 2 (on the right)

The quality of the computed restorations is depicted in Fig. 11. The curves show the RRE when terminating the iterations with the discrepancy principle. We can notice a stable behavior when ρ_{struct} decreases. For Test 1, $\rho \in [10^{-5}, 10^{-2}]$ gives exactly the same results in terms of RRE and number of iterations. Similarly, for Test 2 the results are not sensitive to a decease in ρ. Therefore, a careful tuning of the parameters ρ and ρ_{struct} is not required when using F-GMRES.

4.5 Robustness Analysis of the New Sequence

The sequence (27) depends on the choice of the parameter p. This subsection seeks to shed light on how this choice affects the performance of F-GMRES.

Figure 12 displays the parameters α_k for different values of p. Note that for all values of p, the sequence $\alpha_1, \alpha_2, \alpha_3, \ldots$ has the desired behavior: It decreases in the first few iterations, when little regularization is required, and increases in subsequent iterations when more regularization is needed. It can be seen that for larger values of

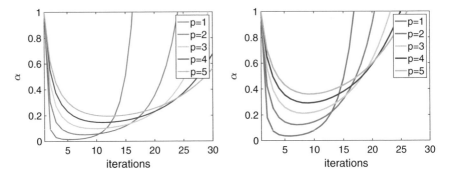

Fig. 12 Plots of α_k for F-GMRES New as a function of the number of iterations k for different values of p for Test 1 (on the left) and for Test 2 (on the right)

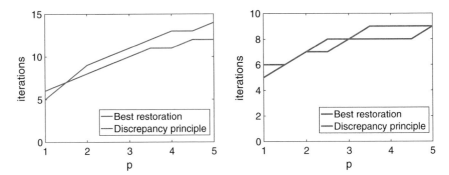

Fig. 13 Plot of the number of iterations required to compute the best restoration by F-GMRES New (blue curve) and plot of the number of iterations required by F-GMRES New when using the discrepancy principle (red curve) as a function of p for Test 1 (on the left) and for Test 2 (on the right)

p, the sequence $\{\alpha_k\}$ first decreases slower and then increases slower than for smaller values of p. Figure 13 displays how p affects the number of iterations required to satisfy the discrepancy principle and to determine the best restoration. We can see that more iterations are required for larger values of p. Moreover, Fig. 13 shows that termination of the iterations with the discrepancy principle works well in the sense that the computed restorations are close to the best restorations.

The quality of the computed restorations is depicted in Fig. 14. The blue curves show the RRE for the best restorations determined by F-GMRES for different values of p. These curves are quite insensitive to the choice of p. The red curves show the RRE when terminating the iterations with the discrepancy principle.

In conclusion, the numerical experiments of this section suggest that the value $p = 2$ is appropriate, because this value yields accurate restorations in a small number of iterations.

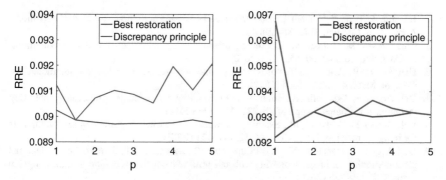

Fig. 14 Plot of the RRE for the best restoration determined by F-GMRES New (blue curve) and the restoration computed by F-GMRES New using the discrepancy principle (red curve) as a function of p for Test 1 (on the left) and for Test 2 (on the right)

5 Conclusion

We have considered image deblurring when the point spread function is non-symmetric and anti-reflective boundary conditions are imposed. The use of standard Krylov subspace methods may require a substantial number of iterations. This can make the restoration of large images expensive. This paper describes a family of non-stationary structure-preserving preconditioners that are designed to reduce the number of iterations. The parameters α_k that define the preconditioners are determined automatically during the iterations. We have focused on the application of these preconditioners in conjunction with the F-GMRES iterative method. Numerical results indicate that this solution approach is competitive with respect to the computational effort required and the quality of the computed restorations.

Acknowledgements We would like to thank the referees and the editor for their valuable comments and suggestions, which helped us to improve the readability and the content of the paper. The first two authors are members of the INdAM Research group GNCS, which has partially supported this work. Research by the third author is supported by NSF grants DMS-1720259 and DMS-1729509. Part of the work of the third author was carried out during a visit to Como. He would like to thank the second author for making this visit possible and enjoyable.

References

1. Aricò, A., Donatelli, M., Nagy, J., Serra–Capizzano, S.: The anti-reflective transform and regularization by filtering. In: Numerical Linear Algebra in Signals, Systems, and Control. Lecture Notes in Electrical Engineering, vol. 80, pp. 1–21. Springer, Berlin (2011)
2. Björck, Å.: Numerical Methods for Least Squares Problems. SIAM, Philadelphia (1996)
3. Buccini, A., Donatelli, M., Reichel, L.: Iterated Tikhonov regularization with a general penalty term. Numer. Linear Algebra Appl., **24**, e2089 (12 pages) (2017)

4. Buccini, A., Park, Y., Reichel, L.: Deblurring methods using antireflective boundary conditions. Appl. Math. Comput. **337**, 386–398 (2018)
5. Christiansen, M., Hanke, M.: Deblurring methods using antireflective boundary conditions. SIAM J. Sci. Comput. **30**, 855–872 (2008)
6. Dell'Acqua, P.: A note on Taylor boundary conditions for accurate image restoration. Adv. Comput. Math. **43**, 1283–1304 (2017)
7. Dell'Acqua, P., Donatelli, M., Estatico, C.: Preconditioners for image restoration by reblurring techniques. J. Comput. Appl. Math. **272**, 313–333 (2014)
8. Dell'Acqua, P., Donatelli, M., Estatico, C., Mazza, M.: Structure preserving preconditioners for image deblurring. J. Sci. Comput. **72**, 147–171 (2017)
9. Dell'Acqua, P., Donatelli, M., Serra Capizzano, S., Sesana, D., Tablino Possio, C.: Optimal preconditioning for image deblurring with anti-reflective boundary conditions. Linear Algebra Appl. **502**, 159–185 (2016)
10. Dell'Acqua, P., Durastante, F.: New periodicontinuous boundary conditions for fast and accurate image restoration, **43**, 1283–1304 (2017)
11. Donatelli, M., Estatico, C., Martinelli, A., Serra-Capizzano, S.: Improved image deblurring with anti-reflective boundary conditions and re-blurring. Inverse Probl. **22**, 2035–2053 (2006)
12. Donatelli, M., Hanke, M.: Fast nonstationary preconditioned iterative methods for ill-posed problems with application to image deblurring. Inverse Probl. **29**, 095008 (16 pages) (2013)
13. Donatelli, M., Estatico, C., Nagy, J., Perrone, L., Serra–Capizzano, S.: Anti-reflective boundary conditions and fast 2D deblurring models. In: Luk, F.T. (ed.), Advanced Signal Processing Algorithms, Architectures, and Implementations XIII. Proceedings of the SPIE, vol. 5205, pp. 380–389 (2003)
14. Donatelli, M., Martin, D., Reichel, L.: Arnoldi methods for image deblurring with anti-reflective boundary conditions. Appl. Math. Comput. **253**, 135–150 (2015)
15. Donatelli, M., Serra–Capizzano, S.: Anti-reflective boundary conditions for deblurring problems. J. Electr. Comput. Eng. **2010**, Article ID 241467 (18 pages) (2010)
16. Dykes, L., Marcellán, F., Reichel, L.: The structure of iterative methods for symmetric linear discrete ill-posed problems. BIT **54**, 129–145 (2014)
17. Dykes, L., Noschese, S., Reichel, L.: Circulant preconditioners for discrete ill-posed Toeplitz systems. Numer. Algorithms **75**, 477–490 (2017)
18. Fan, Y.W., Nagy, J.G.: Synthetic boundary conditions for image deblurring. Linear Algebra Appl. **434**, 2244–2268 (2011)
19. Gazzola, S., Nagy, J.G.: Generalized Arnoldi-Tikhonov method for sparse reconstruction. SIAM J. Sci. Comput. **36**, B225–B247 (2014)
20. Gazzola, S., Noschese, S., Novati, P., Reichel, L.: Arnoldi decomposition, GMRES, and preconditioning for linear discrete ill-posed problems. Appl. Numer. Math. **142**, 102–121 (2019)
21. Gazzola, S., Novati, P., Russo, M.R.: On Krylov projection methods and Tikhonov regularization. Electron. Trans. Numer. Anal. **44**, 83–123 (2015)
22. Hanke, M., Nagy, J.G.: Restoration of atmospherically blurred images by symmetric indefinite conjugate gradient techniques. Inverse Probl. **12**, 157–173 (1996)
23. Hanke, M., Nagy, J., Plemmons, R.: Preconditioned iterative regularization for ill-posed problems. In: Reichel, L., Ruttan, A., Varga, R.S. (eds.), Numerical Linear Algebra, pp. 141–163. de Gruyter, Berlin (1993)
24. Hansen, P.C., Nagy, J., O'Leary, D.P.: Deblurring Images Matrices, Spectra and Filtering. SIAM Publications, Philadelphia (2005)
25. Hearn, T.A., Reichel, L.: Extensions of the Justen-Ramlau blind deconvolution method. Adv. Comput. Math. **39**, 465–491 (2013)
26. Huang, J., Donatelli, M., Chan, R.H.: Nonstationary iterated thresholding algorithms for image deblurring. Inverse Probl. Imaging **7**, 717–736 (2013)
27. Morikuni, K., Reichel, L., Hayami, K.: FGMRES for linear discrete ill-posed problems. Appl. Numer. Math. **75**, 175–187 (2014)
28. Ng, M., Chan, R.H., Tang, W.C.: A fast algorithm for deblurring models with Neumann boundary conditions. SIAM J. Sci. Comput. **21**, 851–866 (1999)

29. Reeves, S.J.: Fast image restoration without boundary artifacts. IEEE Trans. Image Process. **14**, 1448–1453 (2005)
30. Saad, Y.: A flexible inner-outer preconditioned GMRES algorithm. SIAM J. Sci. Comput. **14**, 461–469 (1993)
31. Saad, Y.: Iterative Methods for Sparse Linear Systems, 2nd edn. SIAM, Philadelphia (2003)
32. Saad, Y., Schulz, M.H.: GMRES: a generalized minimal residual method for solving nonsymmetric linear systems. SIAM J. Sci. Stat. Comput. **7**, 856–869 (1986)
33. Serra-Capizzano, S.: A note on anti-reflective boundary conditions and fast deblurring models. SIAM J. Sci. Comput. **25**, 1307–1325 (2003)

Numerical Investigation of the Spectral Distribution of Toeplitz-Function Sequences

Sean Hon and Andy Wathen

Abstract Solving Toeplitz-related systems has been of interest for their ubiquitous applications, particularly in image science and the numerical treatment of differential equations. Extensive study has been carried out for Toeplitz matrices $T_n \in \mathbb{C}^{n \times n}$ as well as Toeplitz-function matrices $h(T_n) \in \mathbb{C}^{n \times n}$, where $h(z)$ is a certain function. Owing to its importance in developing effective preconditioning approaches, their spectral distribution associated with Lebesgue integrable generating functions f has been well investigated. While the spectral result concerning $\{h(T_n)\}_n$ is largely known, such a study is not complete when considering $\{Y_n h(T_n)\}_n$ with $Y_n \in \mathbb{R}^{n \times n}$ being the anti-identity matrix. In this book chapter, we attempt to provide numerical evidence for showing that the eigenvalues of $\{Y_n h(T_n)\}_n$ can be described by a spectral symbol which is precisely identified.

Keywords Toeplitz matrices · Asymptotic spectral distribution · Circulant preconditioners · Hankel matrices

1 Introduction

Solving Toeplitz-related systems has been an important research problem for their crucial applications in computational science and engineering, especially image processing and numerical methods for differential equations. Different fast solvers have been developed for these systems due to their wide-ranging applicability and the computational consideration of applications. A typical example application on imaging

S. Hon (✉)
Department of Mathematics, Hong Kong Baptist University, Kowloon Tong, Hong Kong
e-mail: seanyshon@math.hkbu.edu.hk

A. Wathen
Mathematical Institute, University of Oxford, Radcliffe Observatory Quarter,
Oxford OX2 6GG, UK
e-mail: wathen@maths.ox.ac.uk

© Springer Nature Switzerland AG 2019
M. Donatelli and S. Serra-Capizzano (eds.), *Computational Methods for Inverse Problems in Imaging*, Springer INdAM Series 36,
https://doi.org/10.1007/978-3-030-32882-5_4

that involves solving Toeplitz systems is image restoration and we refer to [12] for detail and more related applications.

Other than the usual Toeplitz systems, preconditioning for Toeplitz-function matrices $h(T_n)$ has been explored recently, where $h(z)$ is an analytic function. For the special case where $h(T_n) \in \mathbb{R}^{n \times n}$ is (real) nonsymmetric, we showed in [11] that one can premultiply it by the anti-identity matrix Y_n defined as

$$
Y_n = \begin{bmatrix} & & 1 \\ & \cdot^{\cdot^{\cdot}} & \\ 1 & & \end{bmatrix} \in \mathbb{R}^{n \times n}
$$

to obtain the symmetrized matrix $Y_n h(T_n)$ without normalizing the original matrix. Provided that a suitable circulant preconditioner $|h(C_n)|$ is used, we also proved that the eigenvalues of $|h(C_n)|^{-1} Y_n h(T_n)$ are clustered around ± 1 under certain assumptions. On a related note, optimal circulant preconditioners were firstly shown to be effective in [8] for several trigonometric functions of Toeplitz matrices. It was further proved in [9, 11] that several common circulant preconditioners can render clustered spectra around ± 1 for $h(T_n)$.

This book chapter is devoted to investigating the asymptotic spectral distribution of $\{Y_n h(T_n[f])\}_n$ for the following reasons.

In the context of iterative solvers for Toeplitz systems, the given matrix $T_n[f]$ is often associated with a generating function f, and it is well-known that the singular value and eigenvalue distributions of $\{T_n[f]\}_n$ can be precisely described by f. Recently, it was shown in [5] that the spectral distribution of $\{Y_n T_n[f]\}_n$ for nonsymmetric $T_n[f]$ generated by complex function $f \in L^1([-\pi, \pi])$ can also be described by certain spectral symbol. Inspired by this recent theoretical advance, it is believed that an analogous result also holds for $\{Y_n h(T_n[f])\}_n$. If such a symbol did exist, one could be able to account for the preconditioning approaches on $Y_n h(T_n[f])$ given in [11] under a unified framework.

Furthermore, in a more theoretical point of view, such a spectral result could advance our understanding of Toeplitz matrix sequences which is of interest in the theory of generalized locally Toeplitz sequences (GLTS) [6].

This chapter is organized as follows. We provide some useful preliminaries on Toeplitz matrices in Sect. 2, followed by a brief discussion of $Y_n h(T_n[f])$ in Sect. 3. Finally, numerical examples are provided in Sect. 4 to support our claim that $\{Y_n h(T_n[f])\}_n$ possesses a spectral symbol which is heuristically identified as $\pm |h \circ f|$.

2   Preliminaries on Toeplitz Matrices

We first present some preliminary results on $\{T_n[f]\}_n$ in this section, which will be useful for studying $\{Y_n h(T_n[f])\}_n$.

Assuming the given Toeplitz matrix $T_n[f]$ is associated with the function f via its Fourier series defined on $[-\pi, \pi]$, we have

$$T_n[f] = \begin{bmatrix} a_0 & a_{-1} & \cdots & a_{-n+2} & a_{-n+1} \\ a_1 & a_0 & a_{-1} & & a_{-n+2} \\ \vdots & a_1 & a_0 & \ddots & \vdots \\ & & \ddots & \ddots & a_{-1} \\ a_{n-2} & & \ddots & \ddots & a_{-1} \\ a_{n-1} & a_{n-2} & \cdots & a_1 & a_0 \end{bmatrix} \in \mathbb{C}^{n \times n},$$

where

$$a_k = \frac{1}{2\pi} \int_{-\pi}^{\pi} f(x) e^{-ikx}\, dx, \qquad k = 0, \pm 1, \pm 2, \ldots,$$

are the Fourier coefficients of f. The function f is called the *generating function/spectral symbol* of $T_n[f]$. If f is real-valued, then $T_n[f]$ is Hermitian for all n. If f is real-valued, nonnegative, and not identically zero almost everywhere, then $T_n[f]$ is Hermitian positive definite for all n. If f is real-valued and even, $T_n[f]$ is (real) symmetric for all n. We refer to [12] for more properties of Toeplitz matrices.

We introduce the following notation and definition under the framework of the GLTS theory [6], before discussing the asymptotic singular value and spectral distributions of $\{T_n[f]\}_n$ associated with f.

Let $C_c(\mathbb{C})$ (or $C_c(\mathbb{R})$) be the space of complex-valued continuous functions defined on $\mathbb{C}$ (or $\mathbb{R}$) with bounded support and let ϕ be a functional, i.e. any function defined on some vector space which takes values in $\mathbb{C}$. Moreover, if $g : D \subset \mathbb{R}^k \to \mathbb{K}$ ($\mathbb{R}$ or $\mathbb{C}$) is a measurable function defined on a set D with $0 < \mu_k(D) < \infty$ where μ_k is the Lebesgue measurable, the functional ϕ_g is denoted such that

$$\phi_g : C_c(\mathbb{K}) \to \mathbb{C} \quad \text{and} \quad \phi_g(F) = \frac{1}{\mu_k(D)} \int_D F\big(g(\mathbf{x}_n)\big)\, d\mathbf{x}_n.$$

Definition 1 ([6, Definition 3.1]) Let $\{A_n\}_n$ be a matrix sequence.

1. We say that $\{A_n\}_n$ has an asymptotic singular value distribution described by a functional $\phi : C_c(\mathbb{R}) \to \mathbb{C}$, and we write $\{A_n\}_n \sim_\sigma \phi$, if

$$\lim_{n\to\infty} \frac{1}{n} \sum_{j=1}^n F\big(\sigma_j(A_n)\big) = \phi(F), \qquad \forall F \in C_c(\mathbb{R}).$$

If $\phi = \phi_{|f|}$ for some measurable $f : D \subset \mathbb{R}^k \to \mathbb{C}$ defined on a set D with $0 < \mu_k(D) < \infty$, we say that $\{A_n\}_n$ has an asymptotic singular value distribution described by f and we write $\{A_n\}_n \sim_\sigma f$.

2. We say that $\{A_n\}_n$ has an asymptotic eigenvalue (or spectral) distribution described by a function $\phi : C_c(\mathbb{R}) \to \mathbb{C}$, and we write $\{A_n\}_n \sim_\lambda \phi$, if

$$\lim_{n \to \infty} \frac{1}{n} \sum_{j=1}^{n} F\big(\lambda_j(A_n)\big) = \phi(F), \qquad \forall F \in C_c(\mathbb{C}).$$

If $\phi = \phi_f$ for some measurable $f : D \subset \mathbb{R}^k \to \mathbb{C}$ defined on a set D with $0 < \mu_k(D) < \infty$, we say that $\{A_n\}_n$ has an asymptotic eigenvalue (or spectral) distribution described by f and we write $\{A_n\}_n \sim_\lambda f$.

First Established in [7], the Szegő theorem that describes the singular value and spectral distributions of Toeplitz matrix sequences has undergone a number of extensions. The theorem was consequently extended by Avram and Parter [1, 13]. Tyrtyshnikov [17] later generalized such a distribution result to the multilevel (p-level) Toeplitz matrices generated by complex-valued $f \in L^1([-\pi, \pi]^p)$. As for the multilevel block Toeplitz matrices generated by a matrix-valued Lebesgue integrable function, Tilli, Serra-Capizzano, and Donatelli also studied their asymptotic spectral behaviour as well as the related preconditioning strategies for example in [4, 14–16].

The generalized Szegő theorem is given as follows:

Theorem 1 (Generalized Szegő theorem [7]) *Suppose* $f \in L^1([-\pi, \pi])$. *Let* $T_n[f]$ *be the Toeplitz matrix generated by* f. *Then*

$$\{T_n[f]\}_n \sim_\sigma f.$$

If moreover f *is real-valued, then*

$$\{T_n[f]\}_n \sim_\lambda f.$$

For Hermitian Toeplitz matrices, more can be said about their spectrum via the following theorem by [3]. This localization result was later refined with strict inequalities in [2], also in preconditioning setting.

Theorem 2 ([2, 3]) *Suppose* $f \in L^1([-\pi, \pi])$ *is real-valued. Let* m_f *and* M_f *be the essential infimum and the essential supremum of* f *on* $[-\pi, \pi]$, *respectively, and let* $T_n[f] \in \mathbb{C}^{n \times n}$ *be the Toeplitz matrix generated by* f. *If* $m_f < M_f$, *then for all* $n > 0$

$$m_f < \lambda_k(T_n[f]) < M_f,$$

where λ_k *is the kth eigenvalue of* $T_n[f]$ *arranged in nondecreasing order. Moreover, if* $m_f \geq 0$, *then* $T_n[f]$ *is Hermitian positive definite for all* n.

For a real Toeplitz matrix $T_n[f]$, one can symmetrize it using a simple reordering trick. Namely, one can first premultiply the matrix by the flip matrix Y_n to obtain the symmetric matrix

$$Y_n T_n[f] = \begin{bmatrix} a_{n-1} & a_{n-2} & \cdots & a_1 & a_0 \\ a_{n-2} & & \cdot\cdot & \cdot\cdot & a_{-1} \\ \vdots & a_1 & a_0 & \cdot\cdot & \vdots \\ a_1 & a_0 & a_{-1} & & a_{-n+2} \\ a_0 & a_{-1} & \cdots & a_{-n+2} & a_{-n+1} \end{bmatrix}.$$

The asymptotic spectral distribution of $\{Y_n T_n[f]\}_n$ was first observed in [10] and then showed precisely in [5]. In effect, the eigenvalues of $\{Y_n T_n[f]\}_n$ are distributed as $\pm|f|$ for complex-valued $f \in L^1([-\pi, \pi])$. Having known such a spectral distribution, the authors in [5] provided that a descriptive result on the eigenvalue of the preconditioned matrix $|C_n|^{-1} Y_n T_n[f]$, where $|C_n|$ is a circulant matrix derived from $T_n[f]$ in a standard way.

Before presenting the precise distribution of $\{Y_n T_n[f]\}_n$, we introduce the following notation. Given $D \subset \mathbb{R}^k$ with $0 < \mu_k(D) < \infty$, we define $\tilde{D}$ as $D \bigcup D_r$, where $r \in \mathbb{R}^k$ and $D_r = r + D$, with the constraint that D and D_r have non-intersecting interior part, i.e. $D° \bigcap D_r° = \emptyset$. Therefore, we have $\mu_k(\tilde{D}) = 2\mu_k(D)$. Given any g defined over D, we define ψ_g over $\tilde{D}$ in the following fashion

$$\psi_g(x) = \begin{cases} g(x), & x \in D, \\ -g(x - r), & x \in D_r, \ x \notin D. \end{cases}$$

Theorem 3 ([5, Theorem 3.2]) *Suppose $f \in L^1([-\pi, \pi])$ with real Fourier coefficients. Let $T_n[f] \in \mathbb{R}^{n\times n}$ be the Toeplitz matrix generated by f and let $Y_n \in \mathbb{R}^{n\times n}$ be the anti-identity matrix. Then*

$$\{Y_n T_n[f]\}_n \sim_\lambda \psi_{|f|}$$

over the domain $\tilde{D}$ with $D = [0, 2\pi]$ and $r = -2\pi$.

Hence, as a direct consequence of Theorem 3, $Y_n T_n[f]$ is in general symmetric indefinite since roughly half of its eigenvalues are negative/positive except possibly for a number of outliers.

Motivated by such a recent distribution result on $\{Y_n T_n[f]\}_n$, we in this work attempt to provide numerical evidence that a similar distribution also holds for $\{Y_n h(T_n[f])\}_n$. It is emphasized that $h(T_n[f])$ is in general not Toeplitz, so Theorem 3 does not straightforwardly apply.

3 Preliminaries on Functions of Toeplitz Matrices

In this section, the preliminaries on analytic functions of Toeplitz matrices are provided. Throughout, we assume that the given function $h(z)$ is analytic with radius of convergence r. Thus, it suffices to consider the following representation of matrix functions via the Taylor series expansion of $h(z)$. Without loss of generality, we choose $\alpha = 0$ in the series representation in order to simplify notation.

Provided that $\rho(T_n) < r$, where $\rho(T_n)$ denotes the spectral radius of T_n, we have

$h(z) = \sum_{k=0}^{\infty} a_k z^k$ with radius of convergence $r = \left(\lim_{k \to \infty} \left| \frac{a_{k+1}}{a_k} \right| \right)^{-1}$. Accordingly,

we have

$$h(T_n) = \sum_{k=0}^{\infty} a_k T_n^k.$$

We give the following definitions of Toeplitz trigonometric function matrices as examples. Note that since the radius of convergence of these trigonometric functions equals to infinity, their corresponding Toeplitz matrix functions are readily defined with no additional conditions required.

Definition 2 For any Toeplitz matrix $T_n \in \mathbb{C}^{n \times n}$,

$$e^{T_n} = I_n + T_n + \frac{1}{2!} T_n^2 + \frac{1}{3!} T_n^3 + \cdots,$$

$$\sin T_n = T_n - \frac{1}{3!} T_n^3 + \frac{1}{5!} T_n^5 - \frac{1}{7!} T_n^7 + \cdots,$$

$$\cos T_n = I_n - \frac{1}{2!} T_n^2 + \frac{1}{4!} T_n^4 - \frac{1}{6!} T_n^6 + \cdots,$$

$$\sinh T_n = T_n + \frac{1}{3!} T_n^3 + \frac{1}{5!} T_n^5 + \frac{1}{7!} T_n^7 + \cdots,$$

and

$$\cosh T_n = I_n + \frac{1}{2!} T_n^2 + \frac{1}{4!} T_n^4 + \frac{1}{6!} T_n^6 + \cdots.$$

In the special case where T_n is a real Toeplitz matrix, it was shown in the following lemma that $Y_n h(T_n)$ is symmetric provided that $\rho(T_n) < r$.

Lemma 1 ([11, Lemma 6]) *Suppose $h(z)$ is an analytic function on $|z| < r$ with radius of convergence r. Let $Y_n \in \mathbb{R}^{n \times n}$ be the anti-identity matrix. If $A_n \in \mathbb{R}^{n \times n}$ with $\rho(A_n) < r$ is (real) persymmetric, i.e. $Y_n A_n = A_n^T Y_n$, then $h(A_n)$ is also (real) persymmetric.*

To end this section, we introduce the following absolute value circulant preconditioner $|h(C_n)|$, which will be used as a preconditioner for $Y_n h(T_n)$ in the next section.

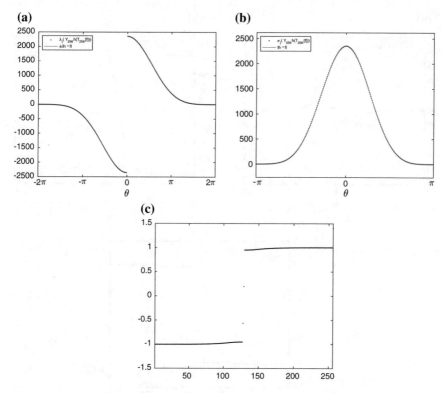

Fig. 1 a Eigenvalues and **b** singular values of $Y_{256}h_1(T_{256}[f])$ when $f(\theta) = 7 + 6\cos\theta$ and $h_1(z) = z^3 + z^2 - z + 1$. **c** Eigenvalues of the preconditioned matrix $|h_1(c(T_{256}[f]))|^{-1}Y_{256}h_1(T_{256}[f])$

Definition 3 ([11]) Suppose $h(z)$ is an analytic function. Let $C_n \in \mathbb{C}^{n \times n}$ be a circulant matrix. The *absolute value circulant matrix* $|h(C_n)| \in \mathbb{C}^{n \times n}$ of $h(C_n)$ is defined by

$$|h(C_n)| = (h(C_n)^* h(C_n))^{1/2} = (h(C_n)h(C_n)^*)^{1/2} = F_n^* |h(\Lambda_n)| F_n,$$

where $F_n \in \mathbb{C}^{n \times n}$ is the Fourier matrix and $|h(\Lambda_n)| \in \mathbb{R}^{n \times n}$ is the diagonal matrix in the eigendecomposition of $h(C_n)$ with all entries replaced by their magnitude.

Remark 1 Due to the well-known diagonalizability of circulant matrices, i.e. $C_n = F_n^* \Lambda_n F_n$, we readily see that both $h(C_n)$ and $|h(C_n)|$ are circulant matrices provided they are well-defined.

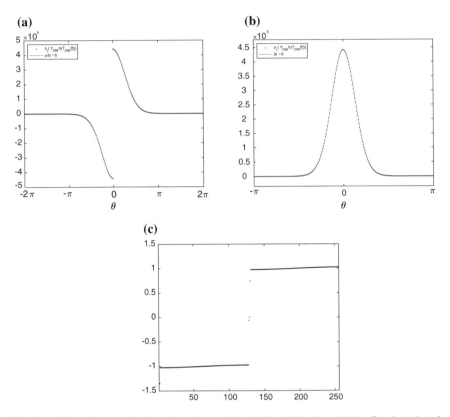

Fig. 2 **a** Eigenvalues and **b** singular values of $Y_{256}h_2(T_{256}[f])$ when $f(\theta) = 7 + 6\cos\theta$ and $h_2(z) = e^z$. **c** Eigenvalues of the preconditioned matrix $|h_2(c(T_{256}[f]))|^{-1}Y_{256}h_2(T_{256}[f])$ (the greatest one is not shown for better visualization)

4 Numerical Experiments

We provide in this section numerical evidence to support our speculation on the spectral distribution of $\{Y_n h(T_n)\}_n$. Note that even though the singular value distribution of $\{Y_n h(T_n)\}_n$ can be obtained as a consequence of the GLTS theory, we still give the related numerical results for completeness.

The analytic functions studied in the tests are as follows:

$$h_1(z) := z^3 + z^2 - z + 1 \quad \text{and} \quad h_2(z) := e^z,$$

and we consider the following four examples concerning different generating functions f.

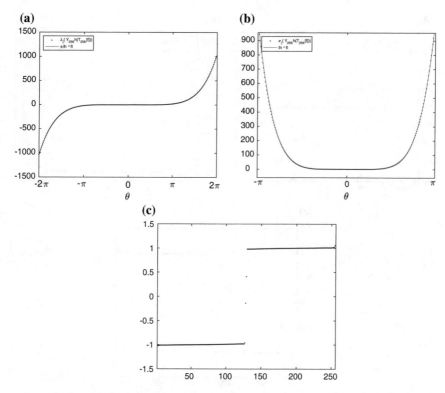

Fig. 3 **a** Eigenvalues and **b** singular values of $Y_{256}h_1(T_{256}[f])$ when $f(\theta) = \theta^2$ and $h_1(z) = z^3 + z^2 - z + 1$. **c** Eigenvalues of the preconditioned matrix $|h_1(c(T_{256}[f]))|^{-1}Y_{256}h_1(T_{256}[f])$ (the greatest one is not shown)

Example 1 We start with the following simple example: the real-valued, even trigonometric polynomial $f : [-\pi, \pi] \mapsto \mathbb{R}$ defined by

$$f(\theta) = 7 + 6\cos\theta,$$

which is periodically extended to the real line. The corresponding Toeplitz matrix $T_n[f] \in \mathbb{C}^{n \times n}$ is the following symmetric positive definite (SPD) tridiagonal matrix:

$$T_n[f] = \begin{bmatrix} 7 & 3 & & \\ 3 & \ddots & \ddots & \\ & \ddots & \ddots & 3 \\ & & 3 & 7 \end{bmatrix}.$$

Example 2 In this example, we consider the real-valued, even trigonometric polynomial $f : [-\pi, \pi] \mapsto \mathbb{R}$ defined by

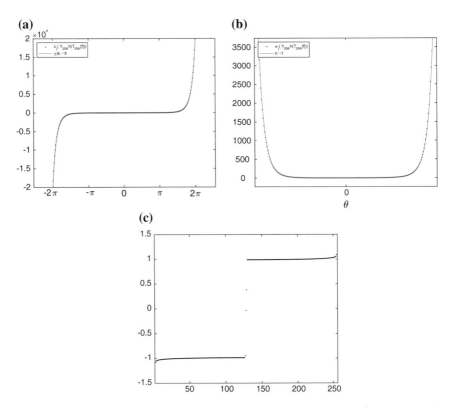

Fig. 4 **a** Eigenvalues and **b** singular values of $Y_{256}h_2(T_{256}[f])$ when $f(\theta) = \theta^2$ and $h_2(z) = e^z$. **c** Eigenvalues of the preconditioned matrix $|h_2(c(T_{256}[f]))|^{-1}Y_{256}h_2(T_{256}[f])$ (the greatest one is not shown)

$$f(\theta) = \theta^2,$$

which is periodically extended to the real line. The corresponding $T_n[f] \in \mathbb{C}^{n \times n}$ is dense and SPD.

Example 3 We now consider the complex-valued, trigonometric polynomial $f :$ $[-\pi, \pi] \mapsto \mathbb{C}$ defined by

$$f(\theta) = 2 + e^{i\theta}.$$

The $n \times n$ Toeplitz matrix generated by f is the following simple nonsymmetric bidiagonal matrix

$$T_n[f] = \begin{bmatrix} 2 & & & \\ 1 & \ddots & & \\ & \ddots & \ddots & \\ & & 1 & 2 \end{bmatrix}.$$

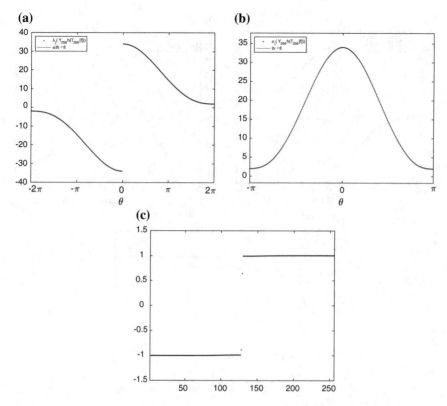

Fig. 5 **a** Eigenvalues and **b** singular values of $Y_{256}h_1(T_{256}[f])$ when $f(\theta) = 2 + e^{i\theta}$ and $h_1(z) = z^3 + z^2 - z + 1$. **c** Eigenvalues of the preconditioned matrix $|h_1(c(T_{256}[f]))|^{-1}Y_{256}h_1(T_{256}[f])$

Example 4 In the last example, we consider the complex-valued, trigonometric polynomial $f : [-\pi, \pi] \mapsto \mathbb{C}$ defined by

$$f(\theta) = \theta^2 + i\theta^3.$$

The Toeplitz matrix generated by f is dense and nonsymmetric.

We first focus on the symmetric matrices in Examples 1 and 2. In Figs. 1, 2, 3 and 4, their distribution results are presented. It is worth noticing that $h(T_n)$ is SPD in these examples, since T_n is SPD by Theorem 2. Hence, we know that the eigenvalues of the symmetric matrix $Y_n h(T_n)$ are the same as those of $h(T_n)$ up to a $(\pm)$ sign using a standard linear algebraic argument. However, as we can see from the figures, the spectral distribution of $\{Y_n h(T_n)\}_n$ appears to have a more structural pattern.

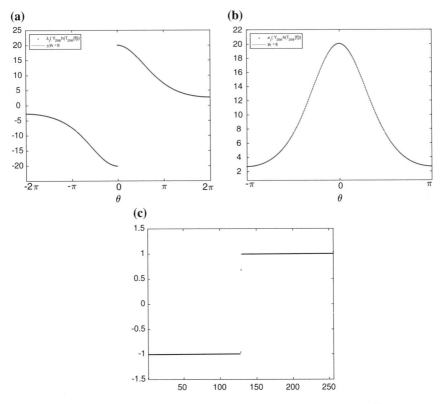

Fig. 6 **a** Eigenvalues and **b** singular values of $Y_{256}h_2(T_{256}[f])$ when $f(\theta) = 2 + e^{i\theta}$ and $h_2(z) = e^z$. **c** Eigenvalues of the preconditioned matrix $|h_2(c(T_{256}[f]))|^{-1}Y_{256}h_2(T_{256}[f])$

We observe from Figs. 1, 2, 3 and 4a that the eigenvalues of $Y_{256}h(T_{256}[f])$ seem to follow the symbol $\psi_{|h\circ f|}$, even though its singular value distribution shown in Figs. 1, 2, 3 and 4b has the expected spectral symbol $|h \circ f|$ which is in accordance with the existing GLTS theory.

We also provide in Figs. 1, 2, 3 and 4c the eigenvalues of the preconditioned matrix $|h(c(T_{256}[f]))|^{-1}Y_{256}h(T_{256}[f])$, where $c(T_{256}[f])$ is the optimal circulant preconditioner derived from $T_{256}[f]$. We refer to [11] for more detail on the related preconditioning results.

While we know the preconditioned matrix has clustered spectra around ± 1 by [11], more about its inertia can be observed from Figs. 1, 2, 3 and 4c—namely there are roughly half eigenvalues are positive/negative. According to Sylvester's law of inertia, both $Y_{256}h(T_{256}[f])$ and $|h(c(T_{256}[f]))|^{-1}Y_{256}h(T_{256}[f])$ share the number of positive, negative, and zero eigenvalues. Thus, these figures provide another evidence to show that the preconditioned matrix sequence could be described by the symbol ψ_1, which suggests the symbol $\psi_{|h\circ f|}$ for $\{Y_n h(T_n)\}_n$.

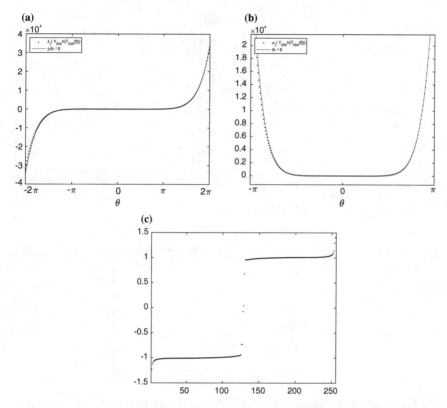

Fig. 7 **a** Eigenvalues and **b** singular values of $Y_{256}h_1(T_{256}[f])$ when $f(\theta) = \theta^2 + i\theta^3$ and $h_1(z) = z^3 + z^2 - z + 1$. **c** Eigenvalues of the preconditioned matrix $|h_1(c(T_{256}[f]))|^{-1}Y_{256}h_1(T_{256}[f])$ (the greatest one is not shown)

At last, we consider the nonsymmetric matrices in Examples 3 and 4. Despite that $h(T_n)$ in these cases is nonsymmetric, the flipped matrix $Y_nh(T_n)$ by Lemma 1 is symmetric. One can again easily deduce that the eigenvalues of $Y_nh(T_n)$ are the same as the moduli of the singular values of $h(T_n)$ up to a ($\pm$) sign. Yet, from Figs. 5, 6, 7 and 8a, we observe a structured spectral distribution for $\{Y_nh(T_n)\}_n$. The corresponding singular values are shown in Figs. 5, 6, 7 and 8b for reference. In Figs. 5, 6, 7 and 8b, we show the eigenvalues of $Y_{256}h(T_{256}[f])$ with $|h(c(T_{256}[f]))|$ as preconditioner. As before, all numerical tests give a good agreement with our speculation.

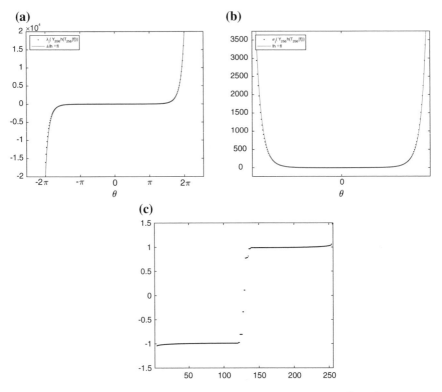

Fig. 8 **a** Eigenvalues and **b** singular values of $Y_{256}h_2(T_{256}[f])$ when $f(\theta) = \theta^2 + i\theta^3$ and $h_2(z) = e^z$. **c** Eigenvalues of the preconditioned matrix $|h_2(c(T_{256}[f]))|^{-1}Y_{256}h_2(T_{256}[f])$

5 Conclusions

We have presented a number of numerical examples to illustrate the asymptotic spectral distribution of $\{Y_n h(T_n)\}_n$. For complex-valued f and certain analytic functions h, the eigenvalues of the symmetrized matrix $Y_n h(T_n[f])$ appear to be effectively described by $\pm|h \circ f|$. These examples suggest that both Theorems 1 and 3 could apply to Toeplitz-function matrix sequences. In a practical point of view, one could also potentially account for the preconditioning strategies used in [11] under a single, coherent framework using such spectral distributions.

References

1. Avram, F.: On bilinear forms in Gaussian random variables and Toeplitz matrices. Probab. Theory Relat. Fields **79**(1), 37–45 (1988)
2. Di Benedetto, F., Fiorentino, G., Serra, S.: C. G. preconditioning for Toeplitz matrices. Comput. Math. Appl. **25**(6), 35–45 (1993)
3. Chan, R.: Toeplitz preconditioners for Toeplitz systems with nonnegative generating functions. IMA J. Numer. Anal. **11**(3), 333–345 (1991)

4. Donatelli, M., Garoni, C., Mazza, M., Serra-Capizzano, S., Sesana, D.: Spectral behavior of preconditioned non-Hermitian multilevel block Toeplitz matrices with matrix-valued symbol. Appl. Math. Comput. **245**, 158–173 (2014)
5. Ferrari, P., Furci, I., Hon, S., Ayman Mursaleen, M., Serra-Capizzano, S.: The eigenvalue distribution of special 2-by-2 block matrix-sequences with applications to the case of symmetrized Toeplitz structures. SIAMJ. Matrix Anal. Appl. **40**(3), 1066–1086 (2019)
6. Garoni, C., Serra Capizzano, S.: Generalized Locally Toeplitz Sequences: Theory and Applications, vol. I. Springer, Cham (2017)
7. Grenander, U., Szegő, G.: Toeplitz Forms and Their Applications, 2nd edn. Chelsea Publishing Co., New York (1984)
8. Hon, S.: Optimal preconditioners for systems defined by functions of Toeplitz matrices. Linear Algebr. Its Appl. **548**, 148–171 (2018)
9. Hon, S.: Circulant preconditioners for functions of Hermitian Toeplitz matrices. J. Comput. Appl. Math. **352**, 328–340 (2019)
10. Hon, S., Ayman Mursaleen, M., Serra-Capizzano, S.: A note on the spectral distribution of symmetrized Toeplitz sequences. Linear Algebr. Its Appl. (2019)
11. Hon, S., Wathen, A.: Circulant preconditioners for analytic functions of Toeplitz matrices. Numer. Algorithms **79**(4), 1211–1230 (2018)
12. Ng, M.: Iterative Methods for Toeplitz Systems. Numerical Mathematics and Scientific Computation. Oxford University Press, New York (2004)
13. Parter, S.: On the distribution of the singular values of Toeplitz matrices. Linear Algebr. Its Appl. **80**, 115–130 (1986)
14. Serra, S.: Asymptotic results on the spectra of block Toeplitz preconditioned matrices. SIAM J. Matrix Anal. Appl. **20**(1), 31–44 (1998)
15. Serra Capizzano, S., Tilli, P.: Extreme singular values and eigenvalues of non-Hermitian block Toeplitz matrices. J. Comput. Appl. Math. **108**(1), 113–130 (1999)
16. Tilli, P.: A note on the spectral distribution of Toeplitz matrices. Linear Multilinear Algebr. **45**(2–3), 147–159 (1998)
17. Tyrtyshnikov, E., Zamarashkin, N.: Spectra of multilevel Toeplitz matrices: advanced theory via simple matrix relationships. Linear Algebr. Its Appl. **270**(1), 15–27 (1998)

The Hough Transform and the Impact of Chronic Leukemia on the Compact Bone Tissue from CT-Images Analysis

Anna Maria Massone, Cristina Campi, Francesco Fiz
and Mauro Carlo Beltrametti

Abstract Computational analysis of X-ray Computed Tomography (CT) images allows the assessment of alteration of bone structure in adult patients with Advanced Chronic Lymphocytic Leukemia (ACLL), and may even offer a powerful tool to assess the development of the disease (prognostic potential). The crucial requirement for this kind of analysis is the application of a pattern recognition method able to accurately segment the intra-bone space in clinical CT images of the human skeleton. Our purpose is to show how this task can be accomplished by a procedure based on the use of the Hough transform technique for special families of algebraic curves. The dataset used for this study is composed of sixteen subjects including eight control subjects, one ACLL survivor, and seven ACLL victims. We apply the Hough transform approach to the set of CT images of appendicular bones for detecting the compact and trabecular bone contours by using ellipses, and we use the computed semi-axes values to infer information on bone alterations in the population affected by ACLL. The effectiveness of this method is proved against ground truth comparison. We show that features depending on the semi-axes values detect a statistically significant difference between the class of control subjects plus the ACLL survivor and the class of ACLL victims.

_____

A. M. Massone (✉) · M. C. Beltrametti
Dipartimento di Matematica, Università di Genova, via Dodecaneso 35, 16146 Genova, Italy
e-mail: massone@dima.unige.it

M. C. Beltrametti
e-mail: beltrametti@dima.unige.it

C. Campi
Dipartimento di Matematica, Tullio Levi-Civita, Università di Padova, Via Trieste 63, 35121
Padova, Italy
e-mail: cristina.campi@unipd.it

F. Fiz
Nuclear Medicine Unit, Department of Radiology, University of Tuebingen,
Hoppe-Seyler-Straße 3, 72076 Tübingen, Germany
e-mail: francesco.fiz.nm@gmail.com

© Springer Nature Switzerland AG 2019
M. Donatelli and S. Serra-Capizzano (eds.), _Computational Methods for
Inverse Problems in Imaging_, Springer INdAM Series 36,
https://doi.org/10.1007/978-3-030-32882-5_5

Keywords X-ray tomography · Image processing · Pattern recognition · Hough transform · Algebraic plane curves

1 Introduction

A computational analysis of a dataset of X-ray Computed Tomography (CT) images recently assessed the presence of alteration of bone structure in adult patients affected by Advanced Chronic Lymphocytic Leukemia (ACLL) [15]. These data (22 ACLL patients and 22 control subjects) have been analyzed by using a dedicated software, based on the use of active contour models [12, 28], to first identify the skeletal border from CT images, and then to segment regions corresponding to trabecular and compact bone (i.e., the two types of osseous tissue that form skeletal bones). The results showed that the whole body skeletal volume is similar in leukemic patients and in control subjects, while ACLL is associated with a significant trabecular bone volume enlargement, which prevails within the appendicular bones, this suggesting that leukemia causes a measurable bone erosion in the appendicular intraosseous space. Further, the degree of skeletal structure alteration displayed a relevant prognostic significance. For these reasons, the assessment of skeletal alterations of compact bone caused by ACLL might be used as a prognostic marker for the prediction of the clinical course of the disease.

The main limitation of the study presented in [15] is two-fold. On the one hand, the segmentation method based on active contours is utilized there just for the identification of the bone outer profile, while, for the recognition of the inner profile, a significantly less reliable heuristic approach based on thresholding is applied. On the other hand, only the bare volume values of the trabecular and compact bone have been considered, without taking into account geometrical aspects of the erosion.

Bone segmentation is an important task in biomedical imaging, and active contours have been widely used as reliable image segmentation methods. The fundamental idea in active contour models is to start with initial closed shapes, i.e., contours, and iteratively change them by applying shrink/expansion operations subject to constraints from a given image. The contour evolution is controlled by the minimization of an energy function. Truc et al. [30] applied several models toward bone segmentation of CT images. Among them, Gradient Vector Flow active contours [32], geometric active contours [20, 33], geodesic active contours [11], Gradient Vector Flow Fast Geometric active contours [22], and Chan–Vese multi-phase active contours without edges [31] have been tested to segment knee bones from CT images. Constructing a graph from an image, the segmentation problem can be alternatively solved by using techniques for graph cuts in graph theory, where a graph cut is the process of partitioning a graph into disjoint sets. Graph cut framework for object segmentation was proposed in [7] and then developed into a large number of extensions based on either iterative parameter re-estimation and learning, multi-scale or hierarchical approaches, and other techniques with a wide range of applications, including medical applications. An exhaustive survey of these developments is given in [6], together with different examples including segmentation of liver and lung lobes in CT volumes. Graph cuts are also used for the segmentation of vertebral bones from volumetric CT images [1]. Since bone structures are characterized by high intensity levels in CT images, their segmentation can also be obtained by using thresholding-based methods, where either a global or local thresholding approach can be followed. An example of this type of techniques is a fully automatic 3D adaptive threshold-ing method [34] that was proposed and tested on CT images of the calcaneus and vertebrae.

In the present paper we apply a pattern recognition method, based on a recent extension of the Hough Transform (HT) concept, able to detect both the trabecular and compact bone contours in CT images. When first introduced, the Hough transform technique was used to detect straight lines in images [17]. It is based on the *point-line duality* as follows: points in a straight line, defined by a linear equation in the image plane $\langle x, y \rangle$ of the form $y = ax + b$, correspond to lines in the parameter space $\langle A, B \rangle$ that intersect in a single point. This point uniquely identifies the coefficients in the equation of the original straight line (analogous procedures to detect circles and ellipses in images have been then introduced in [14]). Further generalizations include Bayesian and fuzzy approaches to the Hough transform [5, 23], in addition to the well known generalized Hough transform [2], which allows the recognition of arbitrary shapes (even composite shapes like cars) by means of pre-set look-up tables (instead of analytic equations) where scale changes, rotations, figure-ground reversals, and reference point translation describing the shape of interest can be accounted for.

Recently, algebraic geometry arguments have been proposed in [4] in order to uti-lize the HT framework for special families of irreducible algebraic plane curves that share the degree, with applications to medical and astronomical images [21]. A fur-ther generalization proposes an iterative approach to the Hough transform technique for piecewise recognition of rather complex anatomical profiles [24].

This paper has two main objectives. The first objective is to evaluate the HT performances in quantifying the trabecular bone volume with respect to the pre-viously employed technique [15]. To do this both techniques are compared with a ground truth given by manual segmentation from expert users. Our second objective is then to infer geometrical information regarding skeletal structure alterations from the recognized curve parameters and show that the set of parameters characterizing the detected curves can effectively provide prognostic information for ACLL. To achieve these goals, and following anatomical considerations, we consider families of ellipses to recognize appendicular bone contours. In particular, in order to study the bone erosion signature, we use as features of interest the ellipses semi-axes.

This paper is organized as follows. In Sect. 2, theoretical and computational de-tails concerning the Hough transform technique, together with a description of the two families of curves here utilized, are presented. In Sect. 3, we provide some infor-mation concerning the patient recruitment and image acquisition details. Section 4 is devoted to achieve the first objective of the paper. Here we show how to address some preprocessing steps through an illustrative example, we apply the recognition technique to the detection of appendicular bone contours, and we offer a quantitative comparison of both, the proposed technique and the one previously employed, with a ground truth. In Sect. 5, we investigate the prognostic significance of the geometrical information inferred by using the HT technique from CT-images for the assessment of bone erosion due to ACLL, second objective of the paper. A brief discussion, together with our conclusions are then offered in Sect. 6. Finally, it is worth noticing that all the tests and analyses presented in this paper were performed within the Matlab computing environment.

2 Background Material

In this section we recall some basic concepts concerning the Hough transform in the case of algebraic plane curves, and we describe two families of curves used in the recognitions presented in Sect. 4. We remark that the general framework here recalled can be exploited for the recognition tasks by using Hough regular families of algebraic curves defined below, including but not limited to straight lines, circles and ellipses. We refer to the first four sections of [25] for a complete, unified exposition on the Hough transform technique with respect to families of curves.

2.1 Hough Transform

We follow the notation introduced in [4, 21]. Let us consider a family of non-constant irreducible real polynomials

$$F(X, Y; \lambda) = \sum_{i,j=0}^{d} g_{ij}(\lambda) X^i Y^j, \quad 0 \leq i + j \leq d, \tag{1}$$

in the variables X, Y, where the coefficients $g_{ij}(\lambda)$ are evaluations in the independent parameters $\lambda = (\lambda_1, \ldots, \lambda_t)$, varying in an Euclidean open set $\mathcal{U} \subseteq \mathbb{R}^t$, of real polynomials $g_{ij}(\Lambda)$ in the variables $\Lambda = (\Lambda_1, \ldots, \Lambda_t)$. We assume that the degree of the polynomials $F(X, Y; \lambda)$ does not depend on λ. Let $\mathcal{F}$ be the corresponding family of zero loci C_λ of $F(X, Y; \lambda)$, and assume that each C_λ is an irreducible real curve in the affine plane $\mathbb{A}^2_{(X,Y)}(\mathbb{R})$, i.e., C_λ is an irreducible curve over the complex field with infinitely many real points in $\mathbb{A}^2_{(X,Y)}(\mathbb{R})$. So we want a family $\mathcal{F} = \{C_\lambda\}$ of irreducible real curves (up to a finite number of isolated points) which share the degree.

If $P = (x_P, y_P)$ is a point of $\mathbb{A}^2_{(X,Y)}(\mathbb{R})$, then the *Hough transform* of P (with respect to the family $\mathcal{F}$) is the algebraic locus $\Gamma_P(\mathcal{F})$ of the affine space $\mathbb{A}^t_{(\Lambda_1, \ldots, \Lambda_t)}(\mathbb{R})$ defined by the equation $\Gamma_P(\Lambda) := F(x_P, y_P; \Lambda) = 0$, where

$$F(x_P, y_P; \Lambda) = \sum_{i,j=0}^{d} g_{ij}(\Lambda) x_P^i y_P^j, \quad 0 \leq i + j \leq d \tag{2}$$

is a real polynomial in the indeterminates $\Lambda = (\Lambda_1, \ldots, \Lambda_t)$. For a general point P in the image space, $\Gamma_P(\mathcal{F})$ is in fact a hypersurface, that is, a $(t-1)$-dimensional locus in the parameter space [26].

The following general facts hold true, as proved in [4, 25].

(a) *The Hough transforms $\Gamma_P(\mathcal{F})$, when P varies on C_λ, all pass through the point λ.*

(b) *Assume that the Hough transforms $\Gamma_P(\mathcal{F})$, when P varies on C_λ, have a point in common other than λ, say λ'. Thus the two curves $C_\lambda, C_{\lambda'}$ coincide.*

(c) (Regularity property) *The following conditions are equivalent*:

  (i) *for all curves $C_\lambda, C_{\lambda'}$ in $\mathcal{F}$, the equality $C_\lambda = C_{\lambda'}$ implies $\lambda = \lambda'$;*
  (ii) *for each curve C_λ in $\mathcal{F}$, one has*

$$\bigcap_{P \in C_\lambda} \Gamma_P(\mathcal{F}) = \lambda.$$

A family $\mathcal{F}$ which meets one of the above equivalent conditions is said to be *Hough regular*.

Condition (c-ii) is easy to be translated into a discrete framework for curves recognition in images: provided that an edge detection process selects in the image a set of points of interest potentially lying on the curve to be recognized, the intersection of their HTs leads to the identification of the parameter set characterizing the curve. Thus, we look for families $\mathcal{F}$ of curves which satisfy the above equivalent conditions. Condition (c-i) provides an effective way to check condition (c-ii). In fact, the equality $C_\lambda = C_{\lambda'}$ is equivalent to $F(X, Y; \lambda) = kF(X, Y; \lambda')$ for some non-zero constant k. This leads to solve a polynomial system, in the variables $\lambda = (\lambda_1, \ldots, \lambda_t)$, $\lambda' = (\lambda'_1, \ldots, \lambda'_t)$, made up of the equations $g_{ij}(\lambda) = kg_{ij}(\lambda')$ for each pair of indices i, j [4]. Therefore, saying that the family $\mathcal{F}$ is Hough regular simply means that such a polynomial system implies $\lambda = \lambda'$.

Based upon the above theoretical result, a recognition algorithm can be implemented as follows. In short, first we apply to the image an edge detection technique to select ν points of interest, $P_1, \ldots, P_\nu$ (see Sect. 4.1 for a detailed description concerning how we deal with this step in the paper). Then, we discretize the parameter space by means of an appropriate number of cells and, for each point of interest P_j, $j = 1, \ldots, \nu$, we compute the Hough transform $\Gamma_{P_j}(\mathcal{F})$ with respect to a fixed family $\mathcal{F}$ of curves. Next, we apply an accumulator function to count how many times each cell in the parameter space is crossed (*voted*) by the computed HTs. Finally, we look for the cell corresponding to the maximum of the accumulator function: the parameter set associated to that cell provides the curve of the family which best approximates the profile of interest in the image.

The recognition algorithm

- Choose a set of points of interest, say P_j, $j = 1, \ldots, \nu$, in the image space by applying an edge detection algorithm
- Consider a discretization in a region $\mathcal{T}$ of the parameter space given by the choice:
  - sampling points $\lambda_{\mathbf{n}} = (\lambda_{1,n_1}, \lambda_{2,n_2}, \ldots, \lambda_{t,n_t})$
  - cells $\mathbf{C_n} := \left\{ \Lambda \in \mathcal{T} \mid \lambda_k \in \left[\lambda_{k,n_k} - \frac{d_k}{2}, \lambda_{k,n_k} + \frac{d_k}{2} \right), \; k = 1, \ldots, t, \; n_k = 1, \ldots, N_k \right\}$, where $\mathbf{n}$ denotes the multi-index $(n_1, n_2, \ldots, n_t)$, d_k the sampling distance and N_k the number of samples with respect to the component k

- Define an *accumulator matrix* $H = (H_{\mathbf{n}})$

$$H_{\mathbf{n}} = H_{n_1, n_2, \cdots, n_t} := \#\{ P_j \mid \Gamma_{P_j}(\mathcal{F}) \cap \mathbf{C_n} \neq \emptyset, \; 1 \leq j \leq \nu \}$$

- Optimize H

$$\mathbf{n}^* := \operatorname{argmax}_{\mathbf{n}} H_{\mathbf{n}}$$

- Identify the set of optimal parameters $\lambda^* := \lambda_{\mathbf{n}^*} = (\lambda_{1,n_1^*}, \lambda_{2,n_2^*}, \ldots, \lambda_{t,n_t^*})$
- Characterize the equation of the seeked curve C_{λ^*}

> Remark I

The computation of the accumulator function and its maximization is the most time-consuming step of the algorithm. Further, it strongly depends on the number of parameters, since the dimension of the domain of this function exactly corresponds to the number of parameters into play. Even though the theory, and the algorithmic aspects, presented in this section hold true in the above general framework, in practice, the computational burden associated to the accumulator function computation and optimization leads to the need of restricting to families of curves depending on a small number of parameters. On the other hand, evidence shows how to be able to control roto-translations and even scaling the variables is a matter of importance; see, for instance, the discussion in Sect. 2.2.2 below. In short, one should be able to study a family of curves of Eq. (1) up to coordinates transformations of type $(X, Y) \mapsto (s_X X + \mu_1 s_Y Y + c_1, s_Y Y + \mu_2 s_X X + c_2)$, where s_X, s_Y are the scaling factors and μ_1, μ_2, c_1, c_2 take care of the roto-translation of the X, Y axes, respectively. This increasing by up to six the number t of the parameters $\lambda = (\lambda_1, \ldots, \lambda_t)$, and then making heavier all computations. Work to establish such a relevant extent is in progress. We also refer to [29] for related results.

> **Remark II**

The robustness of the recognition algorithm in presence of noise has been widely tested in [4], where the Hough transform algorithm showed to be extremely effective in recognizing curves when embedded in a very noisy framework (up to 99% of noise points), and against random perturbations of the location of the points on the curve.

Further, a bound for the number of points of interest to be considered in the curve, i.e. a bound for the number of Hough transforms to be considered for a successful optimization of the accumulator function in the recognition algorithm, is provided in [3]. Such a bound is consequence of geometrical arguments.

2.2 Families of Curves of Interest

We describe two families of curves. The first one is an illustrative example which also shows the extent of the HT framework in detecting curves in images, while the second one plays a crucial role in the paper.

2.2.1 Curve of Lamet

Consider the family $\mathcal{F} = \{C_{a,b}\}$ of curves of degree m of equation $\frac{X^m}{a^m} + \frac{Y^m}{b} = 1$ for positive real numbers a and b, or, in polynomial form (1),

$$C_{a,b} : bX^m + a^m Y^m = a^m b. \qquad (3)$$

Clearly, the curve $C_{a,b}$ is non-singular. The curve of Lamet is bounded for even values of m. Note that, for instance, the case $m = 3$ leads to the unbounded Fermat cubic curve. Indeed (see Example 8 in [25]) the curve of Lamet is contained in the rectangular region

$$\left\{ (x, y) \in \mathbb{A}^2_{(X,Y)}(\mathbb{R}) \mid -a \leq x \leq a, \ -b^{\frac{1}{m}} \leq y \leq b^{\frac{1}{m}} \right\}.$$

For each point $P = (x_P, y_P)$ in the image plane, the HT is the $(m + 1)$-degree curve in the parameter plane $\mathbb{A}^2_{(A,B)}(\mathbb{R})$ of equation

$$\Gamma_P(A, B) : Bx_P^m + A^m y_P^m = A^m B. \qquad (4)$$

Let us assume now $C_{a,b} = C_{a',b'}$. The regularity conditions $g_{ij}(\lambda) = k g_{ij}(\lambda')$, $(i, j) \in \{(m, 0), (0, m), (0, 0)\}$, mentioned before, read in this case $b = kb'$, $a^m = ka'^m$, $-a^m b = -ka'^m b'$ for some $k \in \mathbb{R} \setminus \{0\}$, respectively. Then $ka'^m b' = a^m b =$

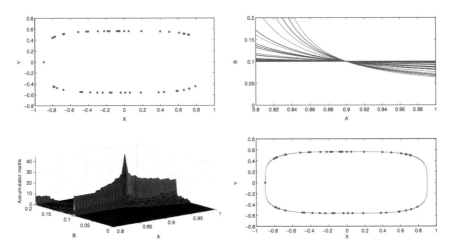

Fig. 1 Recognition of the curve of Lamet, Eq. (3) with $a = 0.9$, $b = 0.1$, $m = 4$. Top left panel: dataset of points randomly sampled on the curve. Top right panel: Hough transforms of the dataset points. Bottom left panel: accumulator matrix. Bottom right panel: recognized curve with dataset points superimposed

$k^2 a'^m b'$, so that $k^2 = k$, whence $k = 1$. Therefore $b = b'$ and $a^m = a'^m$; since $a > 0$ it follows $a = a'$. Thus, the family $\mathcal{F}$ is Hough regular.

Finally, in [3] the optimal bound for the number of Hough transforms to be considered for a successful optimization of the accumulator function is proved to be $\nu_{opt} = m^2 + 1$.

The effectiveness of the recognition algorithm in the case of the curve of Lamet with m fixed to 4 is presented in Fig. 1. For this illustrative example we construct a synthetic database of 50 points ($\nu_{opt} = 17$) satisfying the curve of Eq. (3) with $a = 0.9$ and $b = 0.1$ (see Fig. 1, top left panel). Then the Hough transforms corresponding to all points of the database in the image space are expressed by Eq. (4) and drawn in the top right panel of the figure. In the case we are considering, these transforms are 5-degree curves in the parameter plane $\mathbb{A}^2_{(A,B)}(\mathbb{R})$ that all meet in one point. The accumulator matrix is presented in the bottom left panel. The maximum of this function is clearly visible and it is used in order to determine the parameter values that uniquely identify the curve of Lamet in the image space (Fig. 1, bottom right panel).

2.2.2   Ellipse

To play with ellipses up to roto-translations, it is convenient to consider a family of ellipses expressed in the more general form, as follows. First, look at the general conic of equation $\lambda_0 X^2 + \lambda_1 XY + \lambda_2 Y^2 + \lambda_3 X + \lambda_4 Y + \lambda_5 = 0$. To be an ellipse, we need $\lambda_0 \lambda_2 \neq 0$, and $\lambda_1^2 - 4\lambda_0\lambda_2 < 0$. Thus, we can for instance assume $\lambda_0 = 1$

and look at the 5-parametrized family $\mathcal{F} = \{\mathcal{E}_\lambda\}$ of ellipses expressed in the form

$$\mathcal{E}_\lambda : X^2 + \lambda_1 XY + \lambda_2 Y^2 + \lambda_3 X + \lambda_4 Y + \lambda_5 = 0, \tag{5}$$

where $\lambda_1^2 < 4\lambda_2$ and $\det M \neq 0$, M being the coefficient matrix of Eq. (5). The region $\mathcal{U} \subset \mathbb{R}^5$ where the parameters $\lambda = (\lambda_1, \ldots, \lambda_5)$ vary is then defined by such conditions.

By using the standard invariance and reduction theorems [13], the equation of the ellipse $\mathcal{E}_\lambda$ reduces to the canonical form, with respect to a new system of coordinates $\langle O', X', Y' \rangle$,

$$\mathcal{E}_{a,b} : \frac{X'^2}{a^2} + \frac{Y'^2}{b^2} = 1, \tag{6}$$

where the semi-axes $a, b \in \mathbb{R}_+$ are given by

$$a = \left(-\frac{\det M}{t_1^2 t_2} \right)^{1/2}, \quad b = \left(-\frac{\det M}{t_1 t_2^2} \right)^{1/2}, \tag{7}$$

with t_1, t_2 the eigenvalues of the submatrix $M_{33} = \begin{pmatrix} 1 & \lambda_1/2 \\ \lambda_1/2 & \lambda_2 \end{pmatrix}$. The fact that the conic is an ellipse assures that

$$-\det M/(t_1^2 t_2) > 0, \quad -\det M/(t_1 t_2^2) > 0.$$

Moreover, note that Eq. (6) is defined up to a rotation by an angle of $\pi/2$, which is enough for our purposes.

A straightforward check shows that both the families $\{\mathcal{E}_\lambda\}$ and $\{\mathcal{E}_{a,b}\}$ are Hough regular.

For each point $P = (x_P, y_P)$ in the image plane, the HT of P, with respect to the family $\mathcal{F}$, is the hyperplane in the parameter space $\mathbb{A}^5_{(\Lambda_1,\ldots,\Lambda_5)}(\mathbb{R})$ of equation

$$\Gamma_P(\Lambda) : x_P^2 + \Lambda_1 x_P y_P + \Lambda_2 y_P^2 + \Lambda_3 x_P + \Lambda_4 y_P + \Lambda_5 = 0.$$

Finally, following [3], one has $v_{opt} = d^2 + 1$, where d is the degree of the curves of the family, so that $v_{opt} = 5$ for the ellipses.

3 Patient Recruitment and Image Acquisition

The study presented in this paper is concerned with an analysis of CT-images for the recognition of bone contours of a subset including sixteen subjects among the ones analyzed in [15] (precisely, we have processed eight control subjects belonging to a published normalcy dataset [28], one ACLL survivor, and seven patients, which died because of ACLL). The study presents a retrospective analysis of imaging data, gathered for a valid clinical reason. This analysis was authorized by the Local Ethics Committee (Comitato Etico Regionale Liguria), and influenced in no way the clinical decision making. All patients signed an informed consent prior of study inclusion. Inclusion criteria, as well as imaging technique, have been previously described [15]. Briefly, the study included ACLL patients with no previous specific treatment and recent disease progression. Further exclusion criteria included clinical history of other solid or hematologic malignancy, previous prolonged corticosteroid therapy, previous or ongoing therapy with drugs affecting skeletal metabolism (such as bisphosphonates or denosumab), uncontrolled diabetes, active infection and recent use of erythropoietin, G-CSF or other BM-stimulating drugs. PET/CT imaging started one hour after bolus injection of 18F-fluorodeoxyglucose (FDG, 4.8–5.2 MBq per kilogram of body weight). The exam was performed in the three-dimensional mode, from vertex to toes in an arms-down position, using an integrated PET/CT scanner (Hirez; Siemens Medical Solutions, Knoxville, Tennessee). PET raw data were reconstructed by means of Ordered Subset Expectation Maximization [18] (3 iterations, 16 subsets), and attenuation correction was performed by using CT data. The transaxial field of view and pixel size of the reconstructed PET images were 58.5 cm and 4.57 mm, respectively, with a 128×128 matrix. A 16-detector row helical CT scan was performed with non-diagnostic current and voltage settings, with a gantry rotation speed of 0.5 second and a table speed of 24 mm per gantry rotation. No contrast medium was injected. The entire CT data set was fused with the three-dimensional PET images by using an integrated software interface (Syngo; Siemens, Erlangen, Germany).

4 Image Analysis

The basic step of the analysis presented in this paper is the identification of bone contours in CT data, here performed by using an extended version of the Hough transform recognition algorithm [4, 14, 17]. The Hough transform is widely used in image processing to detect curves (whose equations depend on a set of parameters) in images. The basic idea of this recognition procedure is that points lying on a curve in the image space can be transformed into hypersurfaces (their Hough transforms) in the parameter space, and the set of parameters corresponding to the intersection of all Hough transforms identifies the curve to be recognized in the image space. As shown in Fig. 1, from a computational point of view a histogram (the Hough accumulator) can be defined on the discretized parameter space: for each cell in the parameter space, the value of the accumulator corresponds to the number of Hough transforms passing through that cell. The position of the maximum in the Hough counter identifies the set of parameters characterizing the curve to be detected in the image space.

In order to show how this technique works in the context of this paper, in Sect. 4.1 we present an illustrative example concerning the identification of sternum profiles in the thoracic skeleton. Even though the thoracic skeleton is not included in the following analysis (bone erosion signature has to be found in the appendicular skeleton mainly), this example gives us a chance, first, to show the capability of the HT-based algorithm in detecting a non trivial anatomical profile and, second, to discuss crucial methodological issues like the selection in the image of the points to be HT-transformed. Then, in Sect. 4.2, we show in detail the way we processed the data, and we describe the analyses and comparisons performed in order to evaluate the reliability of our results with respect to a ground truth given by manual segmentation from expert users.

4.1 An Illustrative Example

The sternum is one of the skeleton sites where the bone marrow is abundantly present. Bone marrow is the main hematopoietic tissue in adult humans and its examination can be used for the diagnosis of several diseases, like leukemia. For these reasons, it is important to distinguish between the compact and trabecular bone of this particular district. From a visual inspection of several CT images, it emerges that the sternum profile can be effectively described by using the curve of Lamet, introduced in Sect. 2.2.1 (see [8, 9]). Then, the identification task is performed by means of the HT via curves of Lamet with m fixed to an even value in order to deal with a bounded curve (see Eq. (3)). In particular, we chose the value $m = 4$ proved to be effective in a previous work [8]. Left panel of Fig. 2 shows an axial view of a CT image of a thorax. The white box in the image outlines the sternum region. The result of the

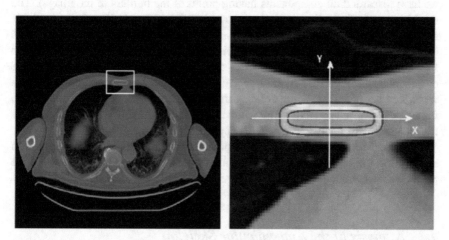

Fig. 2 Recognition of the inner and outer profiles of a sternum. Left panel: Original X-ray CT image with focus on the portion of interest. Right panel: Curves of Lamet (red lines), Eq. (3), providing the best approximation of the outer ($a = 2.025, b = 0.141$) and inner ($a = 1.670, b = 0.011$) sternum profiles superimposed to the original image

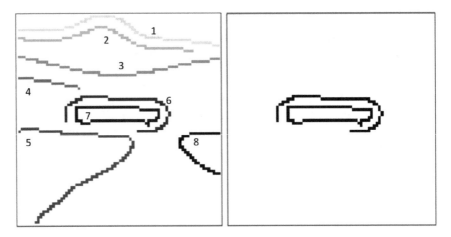

Fig. 3 Extraction of the sternum edge points. Left panel: Application of Canny edge detection to the portion of interest outlined in the left panel of Fig. 2, and resulting identification of eight 8-connected components. Right panel: Rejection of some connected components and extraction of the set of points of interest to be Hough-transformed

identification of the inner and outer bone contours, i.e., the two curves of Lamet (solid lines) associated with the recognized parameters, is given in the right panel.

As a preprocessing step, Canny edge detection [10] (with parameters set to their Matlab default values) followed by a search for connected components [16] in the edge image (Fig. 3, left panel) is used to select the image points of interest, i.e., the ones to be Hough-transformed (due to the anatomical conformation of the sternum, we have discarded all components having points at the borders of the image). The resulting set of points of interest is then shown in the right panel of Fig. 3. A coordinate system (as the one outlined in the right panel of Fig. 2) is then automatically defined in order to have the origin coincident with the center of mass of these identified sternum edge points (the center of mass approximately corresponds to the middle of the trabecular region). No roto-translation with respect to this coordinate system was necessary here. It is worth noticing that the procedure described in this section for the points of interest extraction will be also used in the following subsection for the recognition of bone contours in the appendicular skeleton.

As a final remark we would like to point out the robustness of the method in recognizing the outer profile of the sternum even though a wide portion of edge points was not extracted by the edge detection algorithm.

4.2 Analysis of the Appendicular Skeleton

For each one of the sixteen subjects at disposal, the dataset consists of images from whole-body CT scanning (512×512 pixels per slice, 1.36 mm size each pixel,

Table 1 Minimum and maximum number of slices, corresponding to each appendicular bone in a dataset of images from whole-body CT scanning of sixteen subjects differing for age and sex

	Minimum	Maximum
Left humerus	27	59
Right humerus	25	58
Left femur	38	90
Right femur	36	92

Fig. 4 Appendicular skeleton: 3D (left) and an axial view (right) of the leg bones. In the axial view the compact bone of the femurs (white areas) is clearly visible

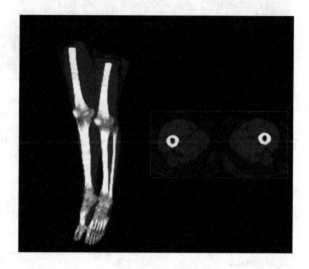

and 5 mm thick images), the number of slices depending on the subjects' height and acquisition modality (for a minimum and maximum number of slices equal to 302 and 476, respectively). Among them we have considered a stack of CT-images corresponding to appendicular bones (i.e., femurs and humeri). For each subject, and for each appendicular bone, we have selected the first and last slice to process, together with a Region Of Interest (ROI) whose dimension was 60×60 pixels. The same ROI was then automatically replicated to each slice in between. Table 1 shows the minimum and maximum number of considered slices, for each appendicular bone, across the population of sixteen subjects.

We applied the Hough transform approach to the whole set of CT images of humeri and femurs for detecting the inner and outer bone contours by using ellipses as prototype curves (see Figs. 4 and 5). To this aim, for each ROI containing either a femur or a humerus, we first selected the points of interest by following the same edge detection-based procedure described in the previous subsection. In Table 2 we report the mean numbers of points of interest, with standard deviations, extracted from the Canny edge detection algorithm for both, the inner and outer bone profile, and for each appendicular bone. Recalling that, in the case of the ellipses, the optimal number of points of interest to guarantee a successful recognition is $\nu_{opt} = 5$, we can conclude

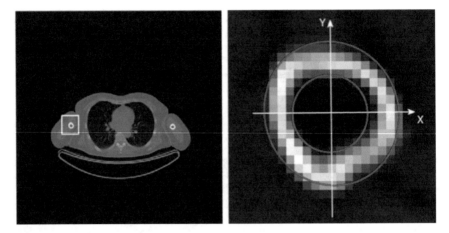

Fig. 5 Recognition of the inner and outer profiles of a humerus. Left panel: Original X-ray CT image with focus on the portion of interest. Right panel: Ellipses, of Eq. (5), providing the best approximation of the outer and inner profiles superimposed to the original image. Outer ellipse: $\lambda_1 = -0.102, \lambda_2 = 0.941, \lambda_3 = 0.014, \lambda_4 = 0.146, \lambda_5 = -2.629$, values leading to $a = 1.7$ and $b = 1.6$ by using relations (7). Inner ellipse: $\lambda_1 = 0, \lambda_2 = 1, \lambda_3 = 0.098, \lambda_4 = 0.144, \lambda_5 = -0.802$, values leading to $a = b = 0.9$ by using relations (7)

Table 2 Mean numbers of points of interest, and corresponding standard deviation values, extracted by the edge detection step from the inner and outer bone contours, for each appendicular bone

	Inner contour	Outer contour
Left humerus	22 ± 6	47 ± 10
Right humerus	23 ± 8	48 ± 10
Left femur	34 ± 11	65 ± 14
Right femur	34 ± 11	65 ± 14

that the cardinality of the datasets at disposal is high enough to guarantee reliable results for all the considered anatomical districts. We introduced a coordinate system centered in the center of mass of the extracted points of interest (i.e., in the trabecular tissue), and finally we used the 5-parametrized family of ellipses $\mathcal{E}_\lambda$ of Eq. (5), which automatically takes into account possible roto-translations. The optimal set of parameters $\lambda = (\lambda_1, \ldots, \lambda_5)$, i.e., the output of the recognition algorithm, was then used to compute the semi-axes values (a and b) by means of relations (7), from which it was also easy to compute the trabecular and compact bone areas (as πab), and the intraosseous volume (by summing up the area values across different slices).

As for the computational time necessary to accomplish the whole process, we used the Matlab *cputime* code to evaluate the CPU time, in seconds, used to run the recognition algorithm on a 2.9 GHz Intel Core i7 processor. Our results show that for a full analysis of a subject about 200 seconds are enough, 120 of which aimed at analysing the outer profiles of both, the right and left femurs.

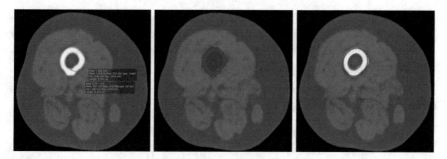

Fig. 6 Computation of the trabecular bone area (A_{in}) and the whole bone area (A_{out}) in a slice of a femur. Left panel: Ground truth, i.e., manually drawn profiles within the OsiriX software package ($A_{\text{in}}^{\text{GT}} = 1.99$ cm^2, $A_{\text{out}}^{\text{GT}} = 5.38$ cm^2). Middle panel: Active contours/thresholding segmentation ($A_{\text{in}}^{\text{ACT}} = 2.31$ cm^2, $A_{\text{out}}^{\text{ACT}} = 5.48$ cm^2). Right panel: Hough transform by using ellipses ($A_{\text{in}}^{\text{HT}} = 1.88$ cm^2, $A_{\text{out}}^{\text{HT}} = 5.43$ cm^2)

4.2.1 Comparison with Ground Truth

As we pointed out in the Introduction, the main limitation of the study presented in [15] lies in the fact that the segmentation method based on active contours was utilized just for the identification of the bone outer profile, while, for the recognition of the inner profile, a less reliable approach based on thresholding was applied. The HT-based approach overcomes this limitation since it is designed to automatically detect both the trabecular and compact bone contours. Taking advantage from this fact, the first aim of this paper is to investigate whether the use of a more accurate detection of the inner profile leads to more reliable results. We have then compared both, the results provided by the HT-based method, and the method based on active contours/thresholding, with a ground truth given by manual segmentation from expert users. Specifically, for each subject, and for each slice of each appendicular bone, the internal and external profiles of the bone were drawn by hand using the OsiriX package [27]; the area within each profile was then automatically computed by the software (see left panel of Fig. 6), and the values compared with the ones obtained by using both the active contours/thresholding model (see middle panel of Fig. 6) and the Hough transform (see right panel of Fig. 6).

For each subject s, $s = 1, \ldots, 16$, we introduce the following notation:

- $i \in \{1, 2, 3, 4\}$, an index identifying the left and right humeri and the left and right femurs, in the order;
- $n(s, i)$, the number of images at disposal corresponding to the ith appendicular bone, as labelled in the previous item;
- $j = 1, \ldots, n(s, i)$, an index identifying the jth image of the ith appendicular bone;
- $A_{\text{in}}^{\text{GT}}(s, i, j)$, $A_{\text{in}}^{\text{ACT}}(s, i, j)$, and $A_{\text{in}}^{\text{HT}}(s, i, j)$, the trabecular bone areas (i.e, inner areas) computed in the jth image of the ith appendicular bone, by using OsiriX (ground truth), the active contours/thresholding method, and the Hough transform technique, respectively;

- $A_{\text{out}}^{\text{GT}}(s, i, j)$, $A_{\text{out}}^{\text{ACT}}(s, i, j)$, and $A_{\text{out}}^{\text{HT}}(s, i, j)$, the whole (trabecular plus compact bone) areas (i.e., outer areas) computed in the jth image of the ith appendicular bone, by using OsiriX (ground truth), the active contours/thresholding method, and the Hough transform technique, respectively.

For each image, we computed the percentage error originated by the last two methods in the evaluation of the trabecular bone area with respect to the ground truth value:

$$err_{\text{in}}^{\text{ACT}}(s, i, j) = \frac{A_{\text{in}}^{\text{ACT}}(s, i, j) - A_{\text{in}}^{\text{GT}}(s, i, j)}{A_{\text{in}}^{\text{GT}}(s, i, j)} \tag{8}$$

$$err_{\text{in}}^{\text{HT}}(s, i, j) = \frac{A_{\text{in}}^{\text{HT}}(s, i, j) - A_{\text{in}}^{\text{GT}}(s, i, j)}{A_{\text{in}}^{\text{GT}}(s, i, j)}, \tag{9}$$

and, analogously, for the whole bone areas:

$$err_{\text{out}}^{\text{ACT}}(s, i, j) = \frac{A_{\text{out}}^{\text{ACT}}(s, i, j) - A_{\text{out}}^{\text{GT}}(s, i, j)}{A_{\text{out}}^{\text{GT}}(s, i, j)} \tag{10}$$

$$err_{\text{out}}^{\text{HT}}(s, i, j) = \frac{A_{\text{out}}^{\text{HT}}(s, i, j) - A_{\text{out}}^{\text{GT}}(s, i, j)}{A_{\text{out}}^{\text{GT}}(s, i, j)}. \tag{11}$$

Note that in relations (8)–(11), we intentionally avoided the use of the absolute values to account for possible underestimations or overestimations of the quantity into play. Further, the computed values can be analyzed in many different ways by averaging over different indices. Here we offer an overall comparison by defining, for each method used and with obvious meaning of the notations, the overall mean percentage errors as:

$$Err_{\text{in}}^{\text{ACT}} = \frac{1}{K} \sum_{s} \sum_{i} \frac{1}{n(s, i)} \sum_{j=1}^{n(s,i)} err_{\text{in}}^{\text{ACT}}(s, i, j) \tag{12}$$

$$Err_{\text{out}}^{\text{ACT}} = \frac{1}{K} \sum_{s} \sum_{i} \frac{1}{n(s, i)} \sum_{j=1}^{n(s,i)} err_{\text{out}}^{\text{ACT}}(s, i, j) \tag{13}$$

$$Err_{\text{in}}^{\text{HT}} = \frac{1}{K} \sum_{s} \sum_{i} \frac{1}{n(s, i)} \sum_{j=1}^{n(s,i)} err_{\text{in}}^{\text{HT}}(s, i, j) \tag{14}$$

$$Err_{\text{out}}^{\text{HT}} = \frac{1}{K} \sum_{s} \sum_{i} \frac{1}{n(s, i)} \sum_{j=1}^{n(s,i)} err_{\text{out}}^{\text{HT}}(s, i, j), \tag{15}$$

Table 3 Overall mean percentage errors, and corresponding standard deviations, computed by using Eqs. (12–15)

Active contours/thresholding	Hough transform
$Err_{in}^{ACT} = (0.1 \pm 0.2)$	$Err_{in}^{HT} = (-0.06 \pm 0.09)$
$Err_{out}^{ACT} = (-0.1 \pm 0.1)$	$Err_{out}^{HT} = (-0.01 \pm 0.08)$

where $K = 16 \times 4$ is a constant allowing one to average over the total number of subjects and appendicular bones. The four computed values, together with the corresponding standard deviations, are summarized in Table 3, where we can appreciate how the active contours/thresholding method overestimates the trabecular bone volume with respect to the ground truth, coherently with a non-optimal performance of the thresholding step in the bone segmentation process. The other three values seem to slightly underestimate the corresponding quantities. All values look very accurate within their standard deviations, even though the HT approach shows, as expected, a more precise behaviour with error values smaller of an order of magnitude than the active contours/thresholding approach. An analogous analysis performed by dividing the population in two classes (control subjects plus the ACLL survivor, ACLL victims) provided very similar results.

5 Prognostic Significance

The results presented in [15], and briefly summarized in Sect. 1, show that the effects of ACLL on skeletal structure cause a significant expansion of the intraosseous volume (in particular in the appendicular skeleton) and no significant alterations in the skeletal bone volume, i.e., ACLL erodes the compact bone from within the trabecular tissue.

In our framework, this evidence translates in the fact that no significant alterations are expected on the semi-axes values associated to the outer ellipses, over the whole population. More importantly, *in the high-risk population, alterations in the compact bone due to erosion must be mirrored in increases of the semi-axes values associated to the inner ellipses with respect to the corresponding values in control subjects.* The second aim of this study was then to investigate whether geometrical information related to the inferred semi-axes values can provide prognostic information for ACLL. Specifically, by introducing the following notation:

- $a_{in}(s, i, j)$ and $b_{in}(s, i, j)$, the semi-axes of the inner ellipse recognized in the jth image of the ith appendicular bone for the sth subject,
- $a_{out}(s, i, j)$ and $b_{out}(s, i, j)$, the semi-axes of the outer ellipse recognized in the jth image of the ith appendicular bone for the sth subject,

we expect both the ratios $\frac{a_{in}}{a_{out}}$ and $\frac{b_{in}}{b_{out}}$ to be smaller in control subjects than in leukemic patients, as the indices s, i and j vary.

To investigate the geometrical structure of the femurs and humeri (and the corresponding alterations in the inner semi-axes values in case of disease), we defined the following features:

$$r_a(s, i) = \frac{1}{n(s, i)} \sum_{j=1}^{n(s,i)} \frac{a_{\text{in}}(s, i, j)}{a_{\text{out}}(s, i, j)}, \tag{16}$$

$$r_b(s, i) = \frac{1}{n(s, i)} \sum_{j=1}^{n(s,i)} \frac{b_{\text{in}}(s, i, j)}{b_{\text{out}}(s, i, j)}, \tag{17}$$

$$r_{ab}(s, i) = \frac{1}{n(s, i)} \sum_{j=1}^{n(s,i)} \frac{a_{\text{in}}(s, i, j) b_{\text{in}}(s, i, j)}{a_{\text{out}}(s, i, j) b_{\text{out}}(s, i, j)}, \tag{18}$$

where, for the sth subject and for the ith appendicular bone, $r_a(s, i)$ represents the mean value of the ratio between the inner and outer value of the semi-axis a. Analogously, $r_b(s, i)$ represents the mean value of the ratio between the inner and outer value of the semi-axis b, and eventually $r_{ab}(s, i)$ represents the mean value of the ratio between the areas of the inner and outer ellipses.

Relations (16)–(18) provide information for each subject, and for each appendicular bone. We have processed these information in two different ways.

First, averaging each one of the previous mean ratios over the four anatomical districts (i.e., with respect to the index i) allowed us to identify each subject s by a point $(R_a^{(s)}, R_b^{(s)}, R_{ab}^{(s)}) \in (0, 1) \times (0, 1) \times (0, 1)$, that is, belonging to the space $\mathbb{A}_{(R_a, R_b, R_{ab})}^3(\mathbb{R})$ (see Fig. 7). In this feature space the high-risk subjects are expected to be the points closest to $(1, 1, 1)$. In Fig. 7 we show the distribution of the sixteen subjects with respect to these three features: the survived patient (red circle) falls in the middle of the cloud of the control subjects (green crosses), while the dead patients (black triangles) show a more scattered behavior towards higher values of the three features, as expected. Impact of $R_a^{(s)}$, $R_b^{(s)}$ and $R_{ab}^{(s)}$ values on survival was tested with the Kaplan–Meier [19] analysis. For each parameter R_a, R_b, R_{ab}, the population was stratified in two subgroups, using their median values as a discriminant. For instance, in the case of the R_a parameter, the majority of events occurred in the subgroup with R_a above the median; accordingly, two-years survival was 87.5% and 25% for the subgroups below and above the median, respectively. Figure 8 depicts the survival curve associated with the parameter R_a. Indeed, this analysis produced identical results for the parameters R_b and R_{ab} as well. These data suggest a possible prognostic role of the Hough transform based analysis, which appears to be able to identify a high-risk sub-population, recognized on the base of osseous alterations.

After dividing the population in two classes (control subjects plus the ACLL survivor, ACLL victims), a second way to process information given from relations (16)–(18) is to average the $r_a(s, i)$, $r_b(s, i)$ and $r_{ab}(s, i)$ values over these two sub-populations (i.e., with respect to two disjoint subsets of indices s). In Fig. 9, for each appendicular bone and for each sub-population, we show the resulting mean values, $r_a(i)$, $r_b(i)$ and $r_{ab}(i)$, and the corresponding standard deviations. In non-survivors,

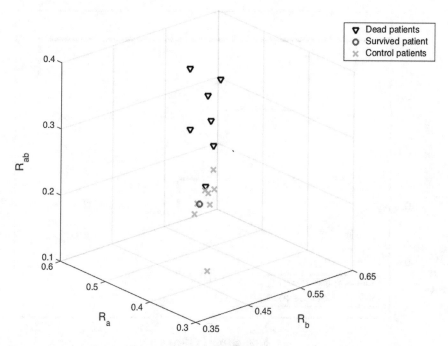

Fig. 7 Distribution of dead patients (black triangles), survived patient (red circle) and controls (green crosses) in terms of the values $R_a^{(s)}$, $R_b^{(s)}$ and $R_{ab}^{(s)}$ in $\mathbb{A}_{(R_a, R_b, R_{ab})}^3(\mathbb{R})$, $s = 1, \ldots, 16$

the four panels clearly show a significant increase of each considered feature together with higher values for the standard deviations. Moreover, we have computed unpaired t-tests to assess whether the data in the two different classes are significantly different from each other. We have found p-values smaller than 0.01 for all the features in both the humeri, while five features out of six in the femurs result in a significance level of 0.05, the sixth one showing a p-value of $\sim$0.09.

> Remarks

1. Although we are looking for variations in the values of the parameters character-izing the inner ellipses, considering the ratios (16)–(18) allows us to normalize and make them comparable within a population of subjects differing for age, sex and duration of the disease.
2. Relations (16) and (17) deal with ratios of linear quantities, while relation (18) is a second-order feature strictly related to the trabecular bone volume evaluation performed in [15].

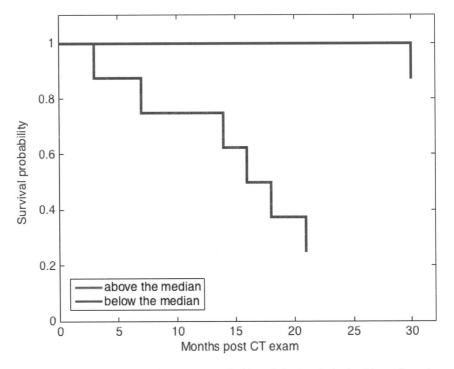

Fig. 8 Kaplan–Meier curves for the two groups of subjects defined on the basis of the median value of their R_a parameter values

6 Discussion and Conclusions

This paper deals with the application of an HT-based approach to the recognition of the compact bone contours in biomedical images, as a tool to assess the impact of Advanced Chronic Lymphocytic Leukemia on the skeleton tissue. We have analyzed CT images of sixteen subjects, seven of which victims of ACLL, and for each appendicular bone we have recognized the inner and outer profiles using ellipses.

As proved in a previous work, ACLL erodes the compact bone from within the trabecular tissue [15]. As a consequence, no significant alterations are expected on the semi-axes values associated to the outer ellipses over the whole population. On the contrary, in the high-risk population alterations of the semi-axes values associated to the inner ellipses with respect to the corresponding values in control subjects are expected. Coherently, the ratios between the inner and outer values for each pair of semi-axes are expected to increase. We have then utilized the parameters characterizing the ellipses, in particular their ratios, to look for findings of bone alteration in the population affected by ACLL compared with the control subjects. To this aim, we have computed first-order and second-order features that all show significant variations in the values of the parameters characterizing the inner ellipses in presence of

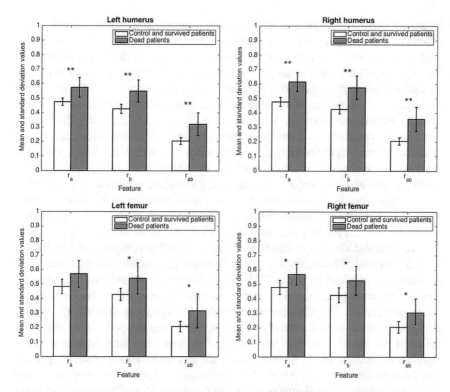

Fig. 9 Measure of the mean values $r_a(i)$, $r_b(i)$, $r_{ab}(i)$ and corresponding standard deviations for two populations: control subjects plus the ACLL survivor (white boxes) and ACLL victims (grey boxes). Left humerus ($i = 1$). Right humerus ($i = 2$). Left femurs ($i = 3$). Right femurs ($i = 4$). The significance levels p (t-test) are also indicated ($** := p \leq 0.01$, and $* := 0.01 < p \leq 0.05$)

disease and then showing for the first time geometrical aspects of the erosion. The different behavior of the two populations, with particular regard to the humeri, is confirmed with a high level of significance by standard statistical tests and Kaplan–Meier analysis. From a quantitative point of view, after stratification according to median values of such features, patients cluster in two groups showing a two-years survival of 87.5% and of 25% according to whether their data are below and above the median, respectively. Further, a t-test analysis, to assess whether the class of control subjects plus the ACLL survivor and the class of ACLL victims are statistically different from each other, showed a significance level $p \leq 0.01$ for the features corresponding to the humeri and $0.01 < p \leq 0.05$ in the femurs case. The method, used in [15] and based on the computational tool validated in [28], was able to provide a previously undisclosed imaging-based prognostic index. The classification outlined in the present paper mirrors and extends previous results, allowing to obtain a more reliable and flexible index to automatically analyze the skeletal composition. This method is able to overcome the main limitation of the approach used in [15], that is, the use of a rather arbitrary thresholding step for the recognition of the (highly informative) bone inner profile. We performed a comparison of both methods with a ground truth given by manual segmentation from expert users, and we have computed the percentage error committed by the two methods in the evaluation of the trabecular bone area with respect to the ground truth value. The result is that the HT-based method significantly outperforms the thresholding procedure, as shown in Table 3. In conclusion, the present study introduces an improved computational method to quantify skeletal alterations in hematologic malignancies. This new technique appears to be reproducible, operator-independent, and could eliminate a number of technical limitations of the formerly proposed active contours/thresholding method, such as the presence of focal interruptions or other disease-related structural alterations, thus allowing for a better characterization and prognostic assessment of hematologic patients. Furthermore, it allows the introduction of the image analysis concept in a patient population where a visual approach of neither morphological nor functional images was able to attain a diagnosis or to formulate a prognosis. Accordingly, further studies should be aimed to test this computational method in several scenarios of hematological or non-hematological-related skeletal alterations, in which the previous approach [28] presented specific limitations, related to the compact bone sampling method or to the inner profile recognition. The limitation of the present study is the relatively small number of subjects at disposal. The present project represents in fact a proof of concept, describing the potential of the application of the Hough transform in a model of leukemia-environment interaction. A rigorous application of the developed method to a large dataset of ACLL patients could assess the diagnostic potential of the features defined in this work and based on the application of the Hough transform to the recognition of bone contours in CT images. To this purpose, it is important to outline how the parameters occurring in the equations defining the curves of the families we deal with, could become concise features for the analysis of the results of the HT-based recognition procedure.

A preliminary version of the Matlab-based software implementing the procedure presented in this paper is fully described in [9] and freely available at the following URL: http://mida.dima.unige.it/g_software_htbone.html.

Acknowledgements We acknowledge support from INDAM grant (intensive period on *Computational Methods for Inverse Problems in Imaging*). We wish to thank Gianmario Sambuceti, Head of the Nuclear Medicine Unit of the IRCCS San Martino - IST, Università degli Studi di Genova, and our colleague and friend Michele Piana for many helpful discussions. Special thanks should be given to Annalisa Perasso for her contribution to the data analysis of this project.

References

1. Aslan, M.S., Ali, A., Rara, H., et al.: A Novel 3D Segmentation of Vertebral Bones From Volumetric CT Images Using Graph Cuts. Advances in Visual Computing, Lecture Notes in Computer Science, vol. 5876, pp. 519–528 (2009)
2. Ballard, D.H.: Generalizing the Hough transform to detect arbitrary shapes. Pattern Recogn. **13**, 111–122 (1981)
3. Beltrametti, M.C., Campi, C., Massone, A.M., Torrente, M.-L.: Geometry of the Hough transforms with applications to synthetic data. arXiv:1904.02587 [cs.CV]
4. Beltrametti, M.C., Massone, A.M., Piana, M.: Hough transform of special classes of curves. SIAM J. Imaging Sci. **6**, 391–412 (2013)
5. Bonci, A., Leo, T., Longhi, S.: A bayesian approach to the Hough transform for line detection. IEEE Trans. Syst. Man. Cybern. Syst.—Part A: Syst. Humans **35**, 945–955 (2005)
6. Boykov, Y., Funka-Lea, G.: Graph cuts and efficient N-D image segmentation. Int. J. Comput. Vis. **70**, 109–131 (2006)
7. Boykov, Y., Jolly, M.-P.: Interactive graph cuts for optimal boundary and region segmentation of objects in N-D images. Int. Conf. Comput. Vis. **I**, 105–112 (2001)
8. Campi, C., Perasso, A., Beltrametti, M.C., Massone, A.M., Sambuceti, G., Piana, M.: Pattern recognition in medical imaging by means of the Hough transform of curves. In: Proceedings of 8th International Symposium on Image and Signal Processing and Analysis (ISPA 2013), pp. 280–283
9. Campi, C., Perasso, A., Beltrametti, M.C., Sambuceti, G., Massone, A.M., Piana, M.: HT-BONE: a graphical user interface for the identification of bone profiles in CT images via extended Hough transform. In: Proceedings of SPIE 9784, Medical Imaging 2016: Image Processing, pp. 978423
10. Canny, J.: A computational approach to edge detection. IEEE Trans. Pattern Anal. Mach. Intell. **8**, 679–698 (1986)
11. Caselles, V., Kimmel, R., Sapiro, G.: Geodesic active contours. Int. J. Comput. Vis. **22**, 61–79 (1997)
12. Chan, T.F., Vese, L.A.: Active contours without edges. IEEE Trans. Image Process. **10**, 266–277 (2001)
13. Ciliberto, C.: Algebra Lineare. Elettronica. Bollati Boringhieri, Programma di Matematica, Fisica (1994)
14. Duda, R.O., Hart, P.E.: Use of the Hough transformation to detect lines and curves in pictures. Commun. ACM. **15**, 11–15 (1972)
15. Fiz, F., Marini, C., Piva, R., et al.: Adult advanced chronic lymphocytic leukemia: computational analysis of whole-body CT documents a bone structure alteration. Radiology **271**, 805–813 (2014)
16. Haralick, R.M., Shapiro, L.G.: Computer and Robot Vision, 1st edn. Addison-Wesley Longman Publishing Co Inc, Boston (1992)
17. Hough, P.V.C.: Method and means for recognizing complex patterns. U.S. Patent 3069654 (1962)
18. Hudson, H.M., Larkin, R.S.: Accelerated image reconstruction using ordered subsets of projection data. IEEE Trans. Med. Imaging **13**, 601–609 (1994)
19. Kaplan, E.L., Meier, P.: Nonparametric estimation from incomplete observations. J. Am. Stat. Assoc. **53**, 457–481 (1958)
20. Malladi, R., Sethian, J.A., Vemuri, B.C.: Shape modeling with front propagation: a level set approach. IEEE Trans. Pattern Anal. Mach. Intell. **17**, 158–175 (1995)
21. Massone, A.M., Perasso, A., Campi, C., et al.: Profile detection in medical and astronomical images by means of the Hough transform of special classes of curves. J. Math. Imaging Vis. **51**, 296–310 (2015)
22. Paragios, N., Mellina-Gottardo, O., Ramesh, V.: Gradient vector flow fast geometric active contours. IEEE Trans. Pattern Anal. Mach. Intell. **26**, 402–407 (2004)

23. Philip, K.P., Dove, E.L., McPherson, D.D., et al.: The fuzzy Hough transform-feature extraction in medical images. IEEE Trans. Med. Imaging **13**, 235–240 (1994)
24. Ricca, G., Beltrametti, M.C., Massone, A.M.: Piecewise recognition of bone skeleton profiles via an iterative Hough transform approach without re-voting. In: Medical Imaging 2015: Image Processing, Proceedings of SPIE, vol. 9413, 94132M:1–8 (2015)
25. Ricca, G., Beltrametti, M.C., Massone, A.M.: Detecting curves of symmetry in images via Hough transform. Math. Comput. Sci., Spec. Issue Geom. Comput. **10**, 179–205 (2016)
26. Robbiano, L.: Hyperplane Sections, Gröbner bases, and Hough transforms. J. Pure Appl. Algebra **219**, 2434–2448 (2015)
27. Rosset, A., Spadola, L., Ratib, O.: OsiriX: An open-source software for navigating in multidimensional DICOM images. J. Digit. Imaging **17**, 205–216 (2004)
28. Sambuceti, G., Brignone, M., Marini, C., et al.: Estimating the whole bone-marrow asset in humans by a computational approach to integrated PET/CT imaging. Eur. J. Nucl. Med. Mol. Imaging **39**, 1326–1338 (2012)
29. Torrente, M.-L., Beltrametti, M.C.: Almost-vanishing polynomials and an application to the Hough transform. J. Algebra Appl. **13**, 39 (2014)
30. Truc, P.T.H., Kim, Y.H., Lee, Y.K., et al.: Evaluation of active contour-based techniques toward bone segmentation from CT images. In: World Congress on Medical Physics and Biomedical Engineering 2006, IFMBE Proceedings, vol. 14, pp. 3121–3125. Springer, Berlin (2007)
31. Vese, L.A., Chan, T.F.: A multiphase level set framework for image segmentation using the Mumford and Shah model. Int. J. Comput. Vis. **50**, 271–293 (2002)
32. Xu, C., Prince, J.L.: Snakes, shapes, and gradient vector flow. IEEE Trans. Image Process. **7**, 359–369 (1998)
33. Yezzi, A., Kichenassamy, S., Kumar, A., et al.: A geometric snake model for segmentation of medical imagery. IEEE Trans. Med. Imaging **16**, 199–209 (1997)
34. Zhang, J., Yan, C.H., Chui, C.K., et al.: Fast segmentation of bone in CT images using 3D adaptive thresholding. Comput. Biol. Med. **40**, 231–236 (2010)

Multiple Image Deblurring with High Dynamic-Range Poisson Data

Marco Prato, Andrea La Camera, Carmelo Arcidiacono, Patrizia Boccacci and Mario Bertero

Abstract An interesting problem arising in astronomical imaging is the reconstruction of an image with high dynamic range, for example a set of bright point sources superimposed to smooth structures. A few methods have been proposed for dealing with this problem and their performance is not always satisfactory. In this paper we propose a solution based on the representation, already proposed elsewhere, of the image as the sum of a pointwise component and a smooth one, with different regularization for the two components. Our approach is in the framework of Poisson data and to this purpose we need efficient deconvolution methods. Therefore, first we briefly describe the application of the Scaled Gradient Projection (SGP) method to the case of different regularization schemes and subsequently we propose how to apply these methods to the case of multiple image deconvolution of high-dynamic range images, with specific reference to the Fizeau interferometer LBTI of the Large Binocular Telescope (LBT). The efficacy of the proposed methods is illustrated both on simulated images and on real images, observed with LBTI, of the Jovian moon Io. The software is available at http://www.oasis.unimore.it/site/home/software.html.

M. Prato (✉)
Dipartimento di Scienze Fisiche, Informatiche e Matematiche, Università di Modena e Reggio Emilia, Via Campi 213/b, 41125 Modena, Italy
e-mail: marco.prato@unimore.it

A. La Camera · P. Boccacci · M. Bertero
Dipartimento di Informatica, Bioingegneria, Robotica e Ingegneria dei Sistemi (DIBRIS), Università di Genova, Via Dodecaneso 35, 16145 Genova, Italy
e-mail: andrea.lacamera@unige.it

P. Boccacci
e-mail: patrizia.boccacci@unige.it

M. Bertero
e-mail: bertero@disi.unige.it

C. Arcidiacono
Osservatorio Astronomico di Padova, Istituto Nazionale di Astrofisica, vicolo dell'Osservatorio 5, 35122 Padova, Italy
e-mail: carmelo.arcidiacono@inaf.it

© Springer Nature Switzerland AG 2019
M. Donatelli and S. Serra-Capizzano (eds.), *Computational Methods for Inverse Problems in Imaging*, Springer INdAM Series 36,
https://doi.org/10.1007/978-3-030-32882-5_6

Keywords Deconvolution · Numerical optimization · Image reconstruction

1 Introduction

Image deconvolution is a classical inverse and ill-posed problem which was investigated since the dawn of regularization theory. Nowadays there exists a plethora of methods and also some good codes for its solution. Examples for astronomical applications are provided by IDAC [34], MISTRAL [38] and AIDA [33]. In these codes a weighted least-squares approximation to the Poisson data fidelity function is used. This approximation is justified by the fact that data are perturbed by both Poisson and Gaussian noise, and, when the number of counts is large, a Poisson distribution can be approximated by a Gaussian one (for a discussion see [49]).

Therefore the question is: why to propose other methods? The answer is that, as far as we know, all regularized deconvolution methods do not produce satisfactory reconstructions in the case of images with high dynamic range, i.e. images where extremely bright and localized sources are superimposed to fainter and smoothly varying structures. In general the reconstructions are affected by significant ringing artifacts around the bright sources.

In this paper we propose methods for dealing with this difficulty in the framework of multiple deconvolution of images satisfying Poisson statistics. This problem arises, for instance, in the case of images obtained by Fizeau interferometry [8], a particular feature of the Large Binocular Telescope (LBT) [31]. Of the two Fizeau interferometers planned for this unique telescope, one, the so-called LBTI [32], has already produced the first interferometric images [21, 37], while the other, LINC-NIRVANA [29], is currently being commissioned and routinely performs Adaptive Optics observations at LBT Observatory [30]. Our group participated in the deconvolution of the first images of LBTI, showing, for the first time, the possibility of resolving volcanic structures on the surface of the Jovian moon Io from ground based observations [21].

Since we need efficient regularized methods, we first describe the application of the Scaled Gradient Projection (SGP) method to the regularized deconvolution of Poisson data. SGP is a general optimization method for the minimization of differentiable objective functions as proposed in [16]; therefore it can be used when the data fidelity function of Poisson data is regularized with the addition of a differentiable function. As concerns the choice of the scaling, we derive it from the Split-Gradient Method (SGM) proposed in [36]. This algorithm, with different kinds of regularization is already implemented in the software package AIRY[1] [19, 23].

For the deconvolution of high dynamic range images we consider an approach proposed in [25, 28] which consists in assuming that the image to be reconstructed is the sum of two components: a point-wise one corresponding to the point bright sources (stars) and a smooth one representing the underlying structures. Moreover,

---

[1] AIRY can be downloaded from http://www.airyproject.eu.

a different regularization is used for the two components: a sparsity enforcing regularization in terms of ℓ_1 norm in pixel space for the point-wise component and a smoothing regularization for the other one. We call this approach a Multi-Component Method (MCM).

Since the ℓ_1 regularization of the point-wise component is unable to produce reconstructions which localize correctly the point sources, in a subsequent paper [35] we improve the MCM method in the particular case of a single star, with known position, surrounded by an unknown accretion disc. Thanks to the knowledge of the position of the star, in that paper we assume that the point-wise component consists of an array which is zero everywhere except in one given pixel (the star), a very strong constraint on this component; moreover simple ℓ_2 regularization is used for the smooth component and an alternating method is proposed for image reconstruction. This approach provides both a satisfactory estimate of the magnitude of the star and a satisfactory reconstruction of the accretion disc.

Having established the relevance of the knowledge of the location of the bright sources, in this paper we extend the approach by defining a suitable objective function where the unknowns are the intensities of the sources and the smooth component; next we consider the addition of a penalization term depending only on the smooth component; finally, by a suitable extension of SGP to this model, we propose a convergent iterative method for the minimization of this function with respect to the full set of variables,. Therefore only convex and differentiable regularizations of the smooth component are considered. Moreover, besides non-negativity, additional constraints such as the flux value of the complete science object (point-wise plus smooth) can be introduced. The focus is on the case of interferometric imaging, i.e. multiple image deconvolution.

Since, especially in the case of interferometric images, the localization of the bright sources can not be easily derived from the observed images, we also introduce an approach able to overcome this difficulty. We call it a Multi-Step Method (MSM). The first step consists in an SGP based deconvolution of the observed images and the result is used for estimating the localization of the bright sources. This allows to produce a mask which is used as an input of an MCM deconvolution; finally the result of this step is used as a background for a simple non-regularized SGP. The addition of the result of this step to that of the previous one produces the final reconstruction.

In conclusion, the main contributions of our paper are the following:

- The extension of SGP to the regularized deconvolution of multiple images of the same target, with specific applications to Fizeau interferometric images; in addition, differentiable edge-preserving regularization functions, introduced in the existing literature, are considered and discussed.
- Introduction of the multi-component model derived from our approach to the deconvolution of multiple images of targets containing bright spots superimposed to smooth and unknown structures.
- Extension of SGP to the minimization of the objective function derived from the previous model.

- Proposal of MSM for the practical application of the multi-component model to the case of interferometric images. The first step provides an estimate of the localization of the bright sources thanks to a standard SGP deconvolution.

The paper is organized as follows. In Sect. 2 we describe the regularization methods for multiple image deconvolution and we briefly discuss the choice of the parameters appearing in these methods. In Sect. 3 we give the SGP algorithm for multiple image deconvolution both in the standard case and in the case of boundary effect correction; the latter is based on an approach proposed in [6] for single image and in [1] for multiple image deconvolution. Section 4 is devoted to the problem of the reconstruction of high-dynamic images. In Sect. 4.1 we introduce our model based on the knowledge of the location of the bright sources and we propose the appropriate objective function and the corresponding SGP method for its minimization while in Sect. 4.2 we describe the approach based on MSM for the case of interferometric imaging. In Sect. 5 we demonstrate the accuracy provided by these methods by deconvolving both simulated and real images of Io observed in M-band with LBTI [21, 37]. Indeed these images are characterized by bright spots, due to volcanic activity, superimposed to the smooth surface of the moon. We also discuss the possibility of performing photometric analysis on the deconvolved images provided by MSM. Finally in Sect. 6 we summarize the results achieved with our approaches.

All methods are implemented in IDL and the codes are available to the users both in AIRY and in the stand-alone package contained in OASIS.[2]

2 The Regularization Methods

In the case of Poisson data, the data-fidelity function is given by the sum of the Csiszár I-divergences [24, 48], also called generalized Kullback–Leibler divergences or cross-entropies, between the p detected and the corresponding computed images

$$J_0(f; g) = \sum_{j=1}^{p} \sum_{m \in S} \left\{ g_j(m) \ln \frac{g_j(m)}{(A_j f)(m) + b_j(m)} + (A_j f)(m) + b_j(m) - g_j(m) \right\} , \quad (1)$$

where m is a multi-index varying on the pixels of the image and object domain S; the detected images g_j, $(j = 1, ..., p)$ correspond to the same science object f; the backgrounds b_j are the known expected values of the sky emission and A_j the known imaging matrices given by $A_j f = K_j * f$; here K_j is the point spread function (PSF) of the jth image and $*$ denotes convolution product. In general the PSFs are normalized to unit flux. If $p > 1$, then the problem we have in mind is that of p different Fizeau interferometric images obtained with different orientations of the baseline [8]. Except for constants, which do not influence the minimizers, $J_0(f; g)$ is the negative logarithm of the likelihood.

---

[2]http://www.oasis.unimore.it/site/home/software.html.

A regularization of the data fidelity function can be derived from a Bayesian approach; in such a way the negative logarithm of the posterior probability distribution is a function with the following structure [7, 27, 36]

$$J_\beta(f; g) = J_0(f; g) + \beta J_1(f) , \qquad (2)$$

where the regularization function $J_1(f)$ derives from the negative logarithm of a Gibbs prior assumed for the unknown solution and the positive parameter β plays the role of a regularization parameter. Then the Maximum A Posteriori (MAP) estimates of the science object f are the solutions of the minimization problem

$$f_\beta = \arg\min_{f \in \Omega} J_\beta(f; g) , \qquad (3)$$

where Ω is either the non-negative orthant or the set of the non-negative arrays such that

$$\sum_{n \in S} f(n) = c , \quad c = \frac{1}{p} \sum_{j=1}^{p} \sum_{m \in S} [g_j(m) - b_j(m)] , \qquad (4)$$

c being the mean flux of the background subtracted images.

We remark that the function J_0 is non-negative and convex; therefore, if we consider regularizers with the same properties, these properties hold also for J_β. Moreover, some regularizers are strictly convex or the intersection of their null space with the null space of J_0 contains only the null element. Then J_β is strictly convex and the MAP solution is unique [14].

The SGM algorithm [36] is based on a decomposition of the gradient of J_1 with the following structure

$$- \nabla J_1(f) = U_1(f) - V_1(f), \quad U_1(f), V_1(f) \geq 0 , \qquad (5)$$

where U_1, V_1 are suitable non-negative arrays. Then, in its most simple form, it is given by

$$f^{(k+1)} = \frac{f^{(k)}}{p\mathbf{1} + \beta V_1(f^{(k)})} \circ \left(\sum_{j=1}^{p} A_j^T \frac{g_j}{A_j f^{(k)} + b_j} + \beta U_1(f^{(k)}) \right) , \qquad (6)$$

where A_j^T is the transposed matrix, $x \circ y$ denotes the pixel by pixel product of two arrays and the quotient symbol denotes their pixel by pixel quotient. The algorithm reduces to the Richardson-Lucy (RL) algorithm by setting $\beta = 0$. Although in principle any vector $\nabla J_1(f)$ can be decomposed as in (5) (and the decomposition is not unique), there exist several widely used regularization terms for which this splitting follows straightly from the gradient expression [36].

In this paper we consider and implement in our software seven different regularization function: three classical Tikhonov regularizations, the cross entropy regularization and three edge-preserving regularizations: the Hypersurface (HS) regularization, the Markov Random Field (MRF) regularization and the regularization function implemented in the MISTRAL code [38] which will be denoted as MIST. The last three functions contain an additional parameter δ, which will be discussed at the end of this section. The expression of the seven regularization functions and the corresponding functions U_1, V_1 are given in Appendix A.

An important issue in the use of regularization methods is the choice of the regularization parameter β which controls the balance between the two terms of $J_\beta(f; g)$. The choice of this parameter is an old problem widely discussed in the mathematical literature in the case of the regularization of the least-square problem. An account can be found, for instance, in [5, 26].

The case of Poisson noise was not so widely investigated. Criteria for the selection of β are proposed in [2, 9] and numerical simulations demonstrate that they work well when the number of counts is large. In the case of mixed Poisson and Gaussian noise (a noise model introduced for taking into account the read-out-noise in the case of images acquired by a CCD camera [46]), Gaussian noise can be approximated by a suitable Poisson noise in the way suggested in [47], and therefore the cited criteria can be applied also in this case.

However, simulations represent ideal cases. In the real world, when astronomical images are pre-processed for flat field correction, bad pixels removal, background subtraction etc., no rule is available because the pre-processing modifies the statistical properties of the noise. Therefore one can only attempt reconstructions with different values of β using some rules for estimating at least its order of magnitude. It is only possible to say that for the observation of a given science object the value of β depends on the noise level, hence on the integration time: if the integration time increases, then the appropriate value of β decreases.

In the case of images affected by Poisson noise the value of J_0 is approximately independent of g (for sufficiently large number of counts) if f is close to the correct solution; more precisely its value is of the order of $N^2/2$ (see, for instance, [8]). This rule is not satisfied if the images have been pre-processed, but it may give the order of magnitude of the first term in the regularized functional, except when the images are rescaled by the astronomers. Since J_0 depends almost linearly on g, then the same rescaling can be done on its value. The estimated value of J_0 provides a hint on the order of magnitude of β if one can estimate the order of magnitude of J_1 by taking into account parameters deduced from the detected images such as the mean of the pixel values or the mean of the gradient values. Since β provides a balance between the two terms of J_β, then one can do a search around the value of β provided by the quotient J_0/J_1 and look for a solution which could be the best for his purposes.

In addition to β the three edge-preserving regularization functions considered in this paper, namely HS, MRF and MIST defined in the Appendix, contain the additional parameter δ. For simplicity we discuss mainly the case of HS regularization, defined in (36), where δ clearly plays the role of a threshold for the values of the modulus of the gradient of the science object (defined as $|D|$ in (28)). Therefore, the role of this parameter is very important.

Indeed, when δ is small with respect to $|D|$, the above-mentioned regularizer behaves essentially as Total Variation (TV) regularization. This point has been demonstrated numerically in [15]. It is obvious that in such a case, since HS is differentiable, one can use very efficient optimization methods and obtain the same results provided by TV with a much lower computational cost. On the other hand, when δ is greater than $|D|$, the regularizer behaves essentially as a Tikhonov regularization, based on the ℓ_2 norm of the gradient (see (31)). This result follows by a first-order Taylor expansion with respect to $|D|/\delta$. Therefore HS behaves as a sort of interpolation between TV and T-2 regularization and may be free of the undesirable cartoon effects due to TV regularization.

The point is to find a good value of δ, i.e. a good thresholding separating regions with small values of the gradient and regions with high values of the gradient. In practice, since the values of the gradient of the unknown science object are also unknown, for estimating a suitable value of δ one can first compute the mean value δ_{mean} of the modulus of the gradient on the observed image; next perform a search of δ around this mean value in order to find the best value for the user. In such a case, HS regularization behaves essentially as a T-2 regularization in regions where the gradient is smaller than δ_{mean} but behaves as an edge-preserving regularization in regions where the gradient is very large (for instance, in the neighborhood of jumps in the values of the scientific target, as those due to the limb of a planet). In these cases HS provide an improvement with respect to TV since it does not introduce cartoon effects in the smooth regions.

For the deconvolution of images with a moderate dynamic range it may be convenient to take a value of δ slightly higher than δ_{mean}, because of the smoothing effect due to the PSF. On the other hand in the deconvolution of images with very high dynamic range, based on the multi components approach described in the following, it is convenient to take a value of δ smaller than δ_{mean} for the reconstruction of the smooth component, because δ_{mean} is affected by the contribution of the bright and localized sources. We take into account this effect in the case of our numerical simulations as well as in the reconstruction of real images.

3 The Scaled Gradient Projection Method

The convergence of SGM is not guaranteed unless some further sufficient decrease strategies are introduced [36]. Moreover, it is known that the convergence speed is slow.

Starting from the remark that SGM can be written as a scaled gradient method with fixed unitary step-length, i.e.,

$$f^{(k+1)} = f^{(k)} - \frac{f^{(k)}}{p\mathbf{1} + \beta V_1(f^{(k)})} \circ \nabla J_\beta(f^{(k)}; g) , \qquad (7)$$

the SGP method [16] is a natural way to accelerate SGM by introducing variable step-lengths and projections.

In its general form, SGP can be applied to the minimization of any smooth objective function subject to a feasible set Ω on which the projection is fast to compute, as in the case of box (possibly with the addition of an equality) constraint. Feasibility of the iterations and stationarity of the limit points of the sequence are achieved by a projection on the constraints P_Ω and a line-search parameter λ_k automatically detected by means of a monotone Armijo backtracking rule [10], thus resulting in the iteration

$$f^{(k+1)} = f^{(k)} + \lambda_k \left(P_\Omega(f^{(k)} - \alpha_k D_k \circ \nabla J_\beta(f^{(k)}; g)) - f^{(k)} \right), \qquad (8)$$

where

$$D_k = \min \left[L_2, \ \max \left\{ L_1, \ \frac{f^{(k)}}{p\mathbf{1} + \beta V_1(f^{(k)})} \right\} \right] \qquad (9)$$

and

$$P_\Omega(y^{(k)}) \equiv \arg\min_{y \in \Omega} (y - y^{(k)})^T D_k^{-1}(y - y^{(k)}). \qquad (10)$$

The choice of the step-length parameter α_k is the one described e.g. in [41] and is based on the BB rules [3], even if any positive step-length can be exploited in the SGP scheme and generalizations with different strategies might be considered (see e.g. [39, 40]).

The SGP method has been used in both standard [11, 12, 41, 50] and blind [22, 42, 43] deconvolution of astronomical images as an effective accelerated RL algorithm. Under mild conditions on the thresholds L_1, L_2 of the scaling matrix D_k, the sequence generated by SGP converges linearly with respect to the objective function values [13], even if several numerical experiments show that the decrease of the objective function is comparable with state-of-the-art approaches for which quadratic convergence rate has been proved [13, 39].

The SGP method can be easily generalized to account for the boundary effect correction proposed in [1, 6] in the following way:

- extend the object f to a broader array $\bar{S}$ containing the region R whose pixels contribute to the observed image defined on $S \subset R$; if we denote by M_R (resp. M_S) the arrays defined over $\bar{S}$ which are 1 over R (resp. S) and 0 outside, then set

$$\alpha_j(n) = \sum_{m \in \bar{S}} K_j(m - n) M_S(m) , \quad n \in \bar{S} ,$$

$$R = \{ n \in \bar{S} \mid \alpha_j(n) \geq \sigma \ \forall j = 1, .., p \} ,$$

where $\sigma > 0$ is a small quantity;
- extend the images g (resp. the backgrounds b) to $\bar{S}$ by zero padding (resp. defining b in $\bar{S} \setminus S$ as constantly equal to the median value of the background radiation in S);

- define the scaling matrix D_k as

$$D_k = M_R \circ \min \left[L_2, \ \max \left\{ L_1, \frac{f^{(k)}}{\alpha + \beta V_1(f^{(k)})} \right\} \right] , \qquad (11)$$

where $\alpha(n) = \displaystyle\sum_{j=1}^{p} \alpha_j(n)$, $n \in \bar{S}$;

- if the flux conservation is considered, substitute the constraint (4) with

$$\frac{1}{p} \sum_{n \in R} \alpha(n) f(n) = c . \qquad (12)$$

We have implemented SGP in IDL using all the regularization functions described in Appendix A, with or without boundary effect correction; the core is the implementation of the SGP algorithm described above.

The software allows the user to provide its own initialization $f^{(0)}$, otherwise a default constant $N \times N$ image with pixel values equal to c/N^2 is used. In the case of boundary effect correction, the default starting point is an image defined over $\bar{S}$ as

$$f^{(0)}(m) = \begin{cases} (cp)/\left(\sum_{n \in R} \alpha(n)\right), & \text{if } m \in R \\ 0, & \text{if } m \in \bar{S} \setminus R \end{cases} . \qquad (13)$$

SGP iteration must be equipped with one or more stopping rules. In general, in the case of regularization, the iteration must be pushed to convergence. To this purpose we use the following stopping criterion, based on the decrease of the objective function

$$|J_\beta(f^{(k+1)}; g) - J_\beta(f^{(k)}; g)| \leq \nu \, J_\beta(f^{(k+1)}; g) \qquad (14)$$

where ν is a tolerance parameter.

4 High-Dynamic Range Image Deconvolution

In this section we describe our approaches to the problem of deconvolving images with a very high dynamic range.

4.1 The Multi-component Method (MCM)

Deconvolving an image might be a particularly challenging problem when the object consists of very bright sources superimposed to diffuse structures, a common situation in astronomical imaging. Typical artifacts appearing in these cases consist in ringing artifacts around the bright sources, thus leading to an unsatisfactory reconstruction of the underlying diffuse structures and, possibly, also to inaccurate evaluation of the intensities of the bright sources.

A possible way to address this problem is to include this information in the model by assuming that the unknown target f is the sum of a point-wise component f_P and an extended and smooth one f_E, exploiting different regularization strategies for each component.

Since f_P is sparse one could use a sparsity enforcing regularization such as that provided by the ℓ_1 norm [25, 28]; however we prefer a much stronger regularization provided by information on the localization of the bright sources when this information can be derived from the observed images. As concerns f_E one can choose one of the regularizers described in Appendix A: for instance a Tikhonov regularizer if it is known that it is very smooth or an edge-preserving one if it contains edges such as the limb of a planet.

In [35] we consider this model by assuming that the positions of the bright sources are exactly known. We propose an iterative algorithm which, at each iteration, performs an RL iteration on the point-wise component (initialized with a mask which is 1 in the pixels of the sources and 0 elsewhere) and an SGM iteration, with T-0 regularization (see Appendix A for a definition), on the extended component. The specific application is the reconstruction of the jet emitted by a young star with known position. In this paper we assume that the sources are localized inside small regions so that one can construct a mask which is 1 on these regions and 0 elsewhere. No additional regularization is introduced for f_P.

For the application of SGP to this situation, namely known localizations of the bright sources, we define $\mathbb{R}_{\geq 0}^{N \times N}$ as the set of $N \times N$ matrices with non-negative entries and P as the prefixed sub-region of the $N \times N$ region S where the bright sources are located; then the resulting minimization problem becomes

$$\min_{(f_E, f_P) \in \overline{\Omega}} J_\beta(f_E, f_P; g) \equiv J_0(f_E + f_P; g) + \beta J_1(f_E) , \qquad (15)$$

where

$$\overline{\Omega} = \{(f_E, f_P) \in \mathbb{R}_{\geq 0}^{N \times N} \times \mathbb{R}_{\geq 0}^{N \times N} \mid f_P(n) = 0 \ \forall n \in S \backslash P \text{ and } f = f_E + f_P \text{ satisfies (4)}\}. \qquad (16)$$

As concerns implementation, the SGP algorithm is applied to a $N \times N$ matrix f_E' and a N_P vector f_P', where N_P is the number of pixels in P, containing the values of f_P belonging to P. The user is asked to provide a $N \times N$ mask M_P equal to 1 where the point sources are located and 0 elsewhere, from which a N_P vector of indexes i_P is automatically computed in order to track the position of the bright sources within the $N \times N$ array.

The core of the IDL code is a unique SGP deconvolution step in which, given initializations $f_E^{(0)}$, $f_P'^{(0)}$, the two arrays are updated at each iteration according to Algorithm 1, where gradients and scaling matrices are computed according to the objective function in (15). This unique step is required if we wish to apply the flux constraint (4); however, deconvolution without this constraint is also possible.

Algorithm 1 Multi-component SGP

Choose the starting point $(f_E^{(0)}, f_P^{(0)}) \in \overline{\Omega}$, define the mask M_P and set the parameters $\mu, \theta \in (0, 1), 0 < L_1^E \leq L_2^E, 0 < L_1^P \leq L_2^P, 0 < \alpha_{min} < \alpha_{max}, k_{max} \in \mathbb{N}$.

Extract the N_P vector $f_P^{\prime(0)}$ from $f_P^{(0)}$ according to the entries i_P of mask M_P equal to 1.

Set $k = 0$ and CHECK = TRUE.

While (CHECK = TRUE) and ($k \leq k_{max}$) do the following steps:

STEP 1. Choose the parameter $\alpha_k \in [\alpha_{min}, \alpha_{max}]$ and define the $N \times N$ and N_P arrays

$$D_k^E = \min\left[L_2^E, \max\left\{L_1^E, \frac{f_E^{(k)}}{p\mathbf{1} + \beta V_1(f_E^{(k)})}\right\}\right] \quad , \quad D_k^P = \min\left[L_2^P, \max\left\{L_1^P, \frac{f_P^{\prime(k)}}{p\mathbf{1}}\right\}\right]. \tag{17}$$

STEP 2. Compute the $N \times N$ and N_P arrays

$$y_E^{(k)} = f_E^{(k)} - \alpha_k D_k^E \circ \nabla_E J_\beta(f_E^{(k)}, f_P^{\prime(k)}; g) \quad , \quad y_P^{\prime(k)} = f_P^{\prime(k)} - \alpha_k D_k^P \circ \nabla_P^\prime J_\beta(f_E^{(k)}, f_P^{\prime(k)}; g). \tag{18}$$

STEP 3. Compute the projection

$$\pi^{(k)} = P_\Omega([\mathcal{V}(y_E^{(k)}); y_P^{\prime(k)}]), \tag{19}$$

where Ω is the set of $N^2 + N_P$ vectors of non-negative components possibly satisfying (4) and P_Ω is defined in (10), re-size the first N^2 components of $\pi^{(k)}$ in a $N \times N$ matrix $\pi_E^{(k)}$ and define the N_P vector $\pi_P^{\prime(k)}$ equal to the last N_P components of $\pi^{(k)}$.

STEP 4. Compute the descent directions

$$d_E^{(k)} = \pi_E^{(k)} - f_E^{(k)} \quad , \quad d_P^{\prime(k)} = \pi_P^{\prime(k)} - f_P^{\prime(k)}. \tag{20}$$

STEP 5. Backtracking loop: compute the smallest positive integer m such that the inequality

$$J_\beta(f_E^{(k)} + \lambda_k d_E^{(k)}, f_P^{\prime(k)} + \lambda_k d_P^{\prime(k)}; g) \leq J_\beta(f_E^{(k)}, f_P^{\prime(k)}; g) + \mu\lambda_k \nabla J_\beta(f_E^{(k)}, f_P^{\prime(k)}; g) \cdot [\mathcal{V}(d_E^{(k)}); d_P^{\prime(k)}] \tag{21}$$

is satisfied with $\lambda_k = \theta^m$.

STEP 6. Set

$$f_E^{(k+1)} = f_E^{(k)} + \lambda_k d_E^{(k)} \quad , \quad f_P^{\prime(k+1)} = f_P^{\prime(k)} + \lambda_k d_P^{\prime(k)} \tag{22}$$

and $k = k + 1$.

STEP 7. If the stopping criterion is satisfied, then set CHECK = FALSE.

End

Define $f_P^{(k)}$ as a $N \times N$ matrix of zeros with entries of indexes i_P equal to $f_P^{\prime(k)}$.

In order to select the line-search parameter λ_k through the Armijo rule (see Step 5 of Algorithm 1), the current image $f^{(k)}$ is computed by merging $f_P^{\prime(k)}$ into a $N \times N$ matrix $f_P^{(k)}$ and computing the sum $f^{(k)} = f_P^{(k)} + f_E^{(k)}$. The bounds L_1, L_2 for the scaling matrices are chosen separately for both components, while the step-length α_k is computed through the "extended" gradient

$$\nabla J_\beta(f_E^{(k)}, f_P^{\prime(k)}; g) = [\mathcal{V}(\nabla_E J_\beta(f_E^{(k)}, f_P^{\prime(k)}; g)); \nabla_P^\prime J_\beta(f_E^{(k)}, f_P^{\prime(k)}; g)], \tag{23}$$

being $\nabla_E J_\beta$ (resp. $\nabla'_P J_\beta$) the gradient of J_β with respect to the first $N \times N$ (resp. last N_P) variables, $\mathcal{V}(\boldsymbol{h})$ the column vectorization of the array $\boldsymbol{h}$ and $[\boldsymbol{h}; \boldsymbol{h}']$ the column vector obtained concatenating $\boldsymbol{h}$ and $\boldsymbol{h}'$. The parameters $\mu, \theta, \alpha_{\min}, \alpha_{\max}$ are the standard SGP ones (see e.g. [41]). As concerns the initializations, the fluxes of the point-wise objects in $\boldsymbol{f}'^{(0)}_P$ are chosen as those of the corresponding pixels of the background-subtracted observed image $\boldsymbol{g}_1$. The remaining flux of the measured images (i.e., the value obtained by subtracting the flux of $\boldsymbol{f}'^{(0)}_P$ from the total flux c defined in (4)) is then spread on a constant $N \times N$ matrix $\boldsymbol{f}^{(0)}_E$, which represents the starting point for the extended object. However, the user can be freely insert his own initialization arrays.

If the boundary effect correction is included, the following modifications to the algorithm have to be considered:

- region S must be replaced by the $N' \times N'$ region $\bar{S}$ $(N' > N)$;
- the set $\overline{\Omega}$ is defined as

$$\overline{\Omega} = \left\{ (\boldsymbol{f}_E, \boldsymbol{f}_P) \in \mathbb{R}^{N' \times N'}_{\geq 0} \times \mathbb{R}^{N' \times N'}_{\geq 0} \mid \boldsymbol{f}_P(\boldsymbol{n}) = 0 \; \forall \boldsymbol{n} \in \bar{S} \backslash P \text{ and } \boldsymbol{f} = \boldsymbol{f}_E + \boldsymbol{f}_P \text{ satisfies (12)} \right\};$$
(24)

- the constant value of the pixels in $\boldsymbol{f}^{(0)}_E$ is computed according to the constraint (12);
- steps 1 and 3 of Algorithm 1 must be reformulated as follows:

STEP 1. Choose the parameter $\alpha_k \in [\alpha_{\min}, \alpha_{\max}]$ and define the $N' \times N'$ and N_P arrays

$$D^E_k = \boldsymbol{M}_R \circ \min \left[L^E_2, \; \max \left\{ L^E_1, \frac{\boldsymbol{f}^{(k)}_E}{\alpha + \beta \boldsymbol{V}_1(\boldsymbol{f}^{(k)}_E)} \right\} \right], \tag{25}$$

$$D^P_k = \min \left[L^P_2, \; \max \left\{ L^P_1, \frac{\boldsymbol{f}'^{(k)}_P}{\alpha} \right\} \right]. \tag{26}$$

STEP 3. Compute the projection

$$\boldsymbol{\pi}^{(k)} = P_\Omega([\mathcal{V}(\boldsymbol{y}^{(k)}_E); \boldsymbol{y}'^{(k)}_P]), \tag{27}$$

where Ω is the set of $N_R + N_P$ vectors (being N_R the number of pixels in R) of non-negative components possibly satisfying (12) and P_Ω is defined in (10), define $\boldsymbol{\pi}^{(k)}_E$ as a $N' \times N'$ matrix of zeros with pixel values in R equal to the first N_R components of $\boldsymbol{\pi}^{(k)}$ and define the N_P vector $\boldsymbol{\pi}'^{(k)}_P$ equal to the last N_P components of $\boldsymbol{\pi}^{(k)}$.

4.2   The Multi-step Method (MSM)

MCM, as described above, assumes that the positions of the bright point-wise sources are (at least approximately) known. This can be true in particular cases but not always, of course. In particular, in our recent attempts of improving the reconstruction of LBTI images [21, 37] we found that it should be important to know the positions of the bright spots corresponding to hot sources on the surface of Io, but these can not be derived from the interferometric images.

Therefore we developed an approach, which we call a Multi-Step Method (MSM) and we propose for the first time in this paper. It can be briefly described at follows and can be applied to both single- and multi-image deconvolution.

- Step 1—Deconvolve the observed image (or images) with some algorithm, for instance SGP without regularization or with an edge-preserving regularization and a small value of β, for obtaining a sharpening of the image (and a removal of the interferometric fringes in the interferometric case).
- Step 2—Determine the centroids of the bright regions which appear as a result of Step 1 and produce a mask which is one over the centroids, or small regions around the centroids, and zero elsewhere.
- Step 3—Apply MCM to the observed image (images) using the previous mask and a regularizer which looks appropriate to the underlying structure. The output of this step is a reconstruction of this structure.
- Step 4—If we denote as f_E the result of the previous step, then we write the unknown object as $f = h + f_E$ and we can recover h by applying SGP, without regularization, to the observed images; thank to the fact that this algorithms concentrates the solution in a few regions, the result is a reconstruction of the bright sources, with possible artifacts consisting of a few bright pixels external to the domains of the sources. Also in this case the algorithm can be pushed to convergence. In alternative, one can use one of the available regularization algorithms, with a small value of the regularization parameter.

The final result is the sum of the results of Step 3 and Step 4. In the next section we will prove the efficacy of this approach to the reconstruction of Io images in M-band.

5   Numerical Results

Because of our recent activity on LBTI images, in this section we focus on images of the Jovian moon Io in M-band. We mainly consider the application of MSM. In order to verify the accuracy of the reconstructions of the real images we generate simulated LBTI images at M-band [21, 37] of an Io-like object with SNR values comparable to those of the real ones. We select M-band images because in this case the hot spots, due to volcanic activity, are seen as very bright sources over the surface of the moon so that they produce very strong ringing artifacts in the case of standard deconvolution methods.

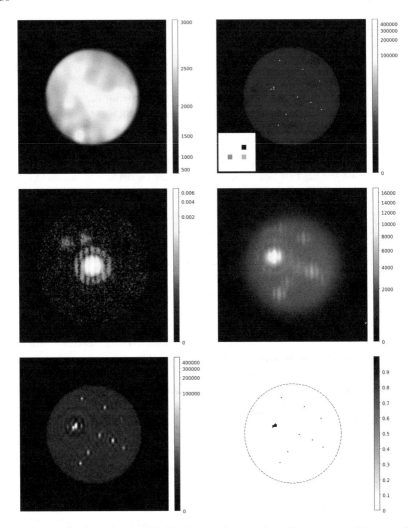

Fig. 1 Simulated and reconstructed Io-like object in the case of complete coverage. *First row—Left*: Model of the surface of the planet (quadratic scale). *Right*: The model after the addition of eleven hot spots (log scale with saturation of the hot spots). We visualize a zoom of the cluster of three spots in the lower left corner of the image. *Second row—Left*: The observed PSF (horizontal baseline), provided with the real images considered in the next section (log scale). *Right*: The corresponding noisy image. *Third row—Left*: The first step of the reconstruction obtained with non-regularized SGP. *Right*: The mask derived from the previous reconstruction. *Forth row—Left*: The reconstructed surface. *Right*: The complete reconstruction (MRF regularization in the third step—see the text)

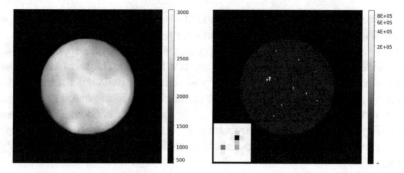

Fig. 1 (continued)

5.1 Simulated Image

In this section we simulate an Io-like object with features similar to those of the M-band LBTI images of Io analysed in the next section. The size of all images is 256×256. As concerns the surface of the moon, we generate a disc with the same diameter of Io in the LBTI images and a smoothly variable brightness, including a sort of limb darkening, and we superimpose bright sources to it. To simulate a structure similar to that of Loki, the dominant structure in the observed LBTI images, we insert a set of three sources which can not be resolved by a 8 m telescope. Moreover we use the PSFs provided with the LBTI images. The simulated surface and the simulated object, obtained by adding to the surface eleven hot, are shown in the first row of Fig. 1 while in the second row we show the PSF and the corresponding image in the case of horizontal baseline of the interferometer.

We first consider a set of four images with orientations of the baseline of 4, 49, 94 and $139°$, thus assuring a good coverage of the u, v plane. In the first step we use a number of iterations of SGP without regularization in order to sharpen the images of the nine hot spots. We compute the centroids of their reconstructions and we find that they coincide with the positions in the original model, except in one case where we find a shift of one pixel. These centroids are used for producing the mask (second step) to be used in the third step. In the third row of Fig. 1 (left panel) we show the result of the first step; it is evident that now the bright spots can be identified but they are encircled by strong ringing effects. In the right panel of the same row the mask obtained by this preliminary reconstruction is also shown.

Finally, in the third step, because of a sharp edge due to the limb of the planet, we decide to test only the three regularizers HS, MRF and MIST. For the three regularizers, we consider a grid of the two parameters δ, β consisting of 11×9 points, with δ varying from 10^{-1} to 10^1 (11 values) and β varying from 10^{-3} to 10^1 (9 values). For each regularizer we consider SGP both with and without flux constraint. For each one of these six cases we find only one minimum of the r.m.s. error, computed as the ℓ_2 norm of the difference between the reconstructed and the original surface of the Io-like object.

The best values of the parameters resulted $\delta = 1$ and $\beta = 10^{-1}$. In the third step we use a tolerance of 10^{-7} but we find that for resolving the Loki-like structure we need a tolerance of 10^{-8} in the stopping criterion for the fourth step, corresponding to about 2500 iterations, a very large number for SGP. The reconstruction with minimal number of point-wise artifacts is obtained by MRF regularization with no flux constraint at the third step and flux constraint at the fourth one. The mean error on the reconstruction of the intensities of the hot points is about 2%. In the last row of Fig. 1 we show the reconstruction of the surface (left panel) and the complete reconstruction of the simulated Io-like object, including the eleven bright spots (right panel).

In a second experiment we consider a situation with a non-complete coverage of the u, v plane, similar to that of the LBTI observations considered in the next section. Therefore the orientations of the baseline correspond to $-30, -22, -16, -8, 4, 16, 29°$ [37]. We use the values of the parameters obtained from the first experiment. Also in this case it is possible to resolve the Loki-like structure with the same tolerances used in the previous experiment but now the required number of SGP iterations at the fourth step is much larger, about 4260.

5.2   Real Image

We consider now the seven interferometric images of Io observed with LBTI during UT 2013 December 24. Observations and data reduction are described in [37]. After pre-processing the images contain negative values as a consequence of background subtraction. These negative values may cause troubles in algorithms which require non-negativity of the observed images. Therefore, instead of zeroing these values, which are small, we add a sufficiently large background such that the average value of the resulting one is not significantly affected by the negative values. We select, arbitrarily, a value of 100 which, of course, is also inserted in the reconstruction algorithm. Three of the observed images (after de-rotation for taking into account the rotation of the baseline with respect to the moon) are shown in the upper panels of Fig. 2. A PSF is also provided, derived from the image of the star HD-78141 and already shown in Fig. 1. During the observation time of about 1 hour the Io relative orbital rotation is of 7.7°. Therefore in the reconstruction of these images we have two additional difficulties: an approximate PSF and a variation of the target during the observation time. We ignore these difficulties in the subsequent analysis.

For the reconstruction of these images we use MRF regularization in the third step. As concerns the values of the parameters β, δ, inspired by the reconstruction of the simulated images with similar SNR values, we considered three values of δ from 0.1 to 1 ($\delta_{mean} = 1.6$) and six values of β from 0.1 to 0.01., with tolerance 10^{-7} for stopping the iteration. By looking at the reconstruction of the surface of the moon, we decided to consider two cases, both with $\delta = 1$: $\beta = 5 \times 10^{-2}$ and $\beta = 10^{-2}$. The corresponding results obtained in the last step by using SGP without regularization

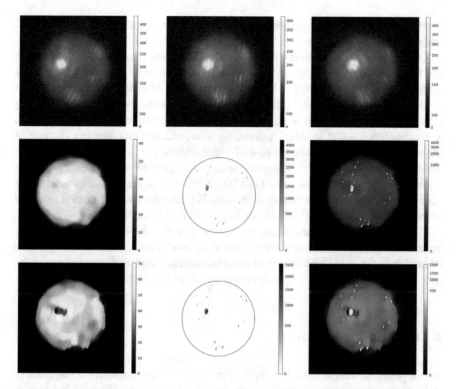

Fig. 2 Reconstruction of the LBTI images of Io at M-band. *First row*—Three interferometric images, showing the variation of the parallactic angle of about 60°. *Second row*—*Left*: The reconstructed surface of Io (linear scale) as obtained at the 3rd step with MRF regularization, $\delta = 1$ and $\beta = 5 \times 10^{-2}$. *Middle*: The reconstruction of the hot spots, in reverse gray sqrt scale (the limb is identified with a dashed circle). *Right*: The complete reconstruction: surface plus hot spots (log scale). *Third row*—Same as in the second row but with $\beta = 10^{-2}$

are shown respectively in the second and third row of Fig. 2. The number of iterations for the last step is 331 in the first case and 113 in the second one.

A general comment is that the reconstruction of the limb is very irregular. This fact is due to two effects: the existence of hot spots close to the limb and a variation of the position of Io inside the image, since the images are centered on Loki, the brightest structure visible in them. A second comment is that the use of a too small regularization parameter does not remove completely ringing artifacts around Loki even if it is able to reduce the point-wise artifacts.

As concerns the eruptions visible in these reconstructions, we do not find evidence for those of Gibil and Rarog (see [21]), even if they were introduced into our mask, consisting of domains of 3×3 pixels for each assumed hot spot (and a broader domain for Loki); we do not exclude their existence but they are certainly very faint so that it is difficult to detect their existence over an estimated and certainly approximated background. Anyway their existence was deduced from reconstructed images full of artifacts.

On the other hand we find two hot spots close to Loki, respectively above right and below left, not considered in the previous reconstructions. If we look at the local reconstructions of Loki, used in [21] and represented in log scale, we find two faint shadows in the same locations; therefore the two hot spots are presumably deriving from these two shadows, as an effect of the last step since non-regularized SGP tends to concentrate the flux in small regions; we also point out that they are stronger in the reconstruction of the second row of Fig. 2 and fainter in that to the third row. Therefore they can be artifacts due to the approximate PSF and the deconvolution method. We must also remark that Loki is so bright that its reconstruction can easily generate artifacts in the deconvolved images. We also observe that Loki is resolved in the reconstruction of the second row but not in that to the third one.

Finally, as concerns photometry, we remark that the eruptions are very well localized in the reconstruction shown in the third row. Since an estimate of the flux of Loki is given in [21], by means of these reconstructions it is easy to compute the ratio between the intensity of one eruption and that of Loki, thus obtaining its flux.

6   Concluding Remarks

In this paper we first discuss the application of SGP to the regularized inversion of Poisson data. Thanks to its efficiency and flexibility, it can be easily used in complex methods we propose for the deconvolution of images with a very high dynamic range. All the methods are implemented in IDL and are at disposal of the reader.

We test the methods for high-dynamic range deconvolution on simulated and real images of the Jovian moon Io. As concerns the real images, we consider LBTI images of a Loki eruption. We show that they provide reconstructions which are free of the usual ringing artifacts even if another kind of artifacts, consisting in a small number of not very bright pixels, is introduced. Anyway these artifacts are not disturbing too much so that it seems that they do not prevent a photometric analysis of the reconstructed[3] images. This point should be very important; however it requires a further and more detailed analysis.

Acknowledgements We thank Al Conrad, LBTO, for permission of using images of Io at M-band, observed with LBTI/LMIRcam [32, 45] during UT 2013 December 24 [21, 37]. Marco Prato is a member of the INdAM Research group GNCS, which is kindly acknowledged.

---

[3]The LBT is an international collaboration among institutions in the United States, Italy and Germany. LBT Corporation partners are: The University of Arizona on behalf of the Arizona Board of Regents; Istituto Nazionale di Astrofisica, Italy; LBT Beteiligungsgesellschaft, Germany, representing the Max-Planck Society, the Leibniz Institute for Astrophysics Potsdam, and Heidelberg University; the Ohio State University, and the Research Corporation, on behalf of the University of Notre Dame, University of Minnesota and University of Virginia.

Appendix

Regularization functions

In this paper we assume that f is a $N \times N$ array extended (when needed) with periodic boundary conditions, i.e., if we set $n = (n_1, n_2)$, then $f(N + 1, n_2) = f(1, n_2)$, $f(n_1, N + 1) = f(n_1, 1)$ and $f(N + 1, N + 1) = f(1, 1)$.

For introducing the regularization functions considered in our methods and software we need some notation. We set $n_{1\pm} = (n_1 \pm 1, n_2)$ and $n_{2\pm} = (n_1, n_2 \pm 1)$ and we introduce the square and the modulus of the discrete gradient

$$D^2(n) = \left[f(n_{1+}) - f(n)\right]^2 + \left[f(n_{2+}) - f(n)\right]^2, \tag{28}$$
$$|D(n)| = \sqrt{\left[f(n_{1+}) - f(n)\right]^2 + \left[f(n_{2+}) - f(n)\right]^2}.$$

Then, the seven regularization functions and the corresponding arrays U_1, V_1 are the following:

- **Zeroth order Tikhonov (T-0) regularization**

$$J_1(f) = \frac{1}{2} \sum_n |f(n)|^2 \ , \tag{29}$$

for which (5) holds by setting

$$U_1(n) = 0 \ , \quad V_1(n) = f(n). \tag{30}$$

- **First order Tikhonov (T-1) regularization**

$$J_1(f) = \frac{1}{2} \sum_n D^2(n) \ , \tag{31}$$

for which (5) holds by setting

$$U_1(n) = \ f(n_{1+}) + f(n_{2+}) + f(n_{1-}) + f(n_{2-}) \ ,$$
$$V_1(n) = \qquad\qquad 4f(n) \ .$$

- **Second order Tikhonov (T-2) regularization**

$$J_1(f) = \frac{1}{2} \sum_n (\Delta f)(n)^2 \ , \tag{32}$$

where Δ denotes the discrete Laplacian. As remarked in [36], it can be written in the form

$$J_1(f) = \frac{1}{2} \sum_n \left[f(n) - (Bf)(n) \right]^2 \ , \tag{33}$$

where B is the convolution matrix obtained from the 3×3 mask with columns $(0, 1/4, 0)$, $(1/4, 0, 1/4)$ and $(0, 1/4, 0)$. Then (5) holds by setting

$$U_1(n) = [(B + B^T)f](n) \ ,$$
$$V_1(n) = [(I + B^T B)f](n) \ .$$

- **Cross-Entropy (CE) regularization** [17, 18]

$$J_1(f) = KL(f, \bar{f}) = \sum_n \left\{ f(n) \ln \left(\frac{f(n)}{\bar{f}(n)} \right) + \bar{f}(n) - f(n) \right\} \ , \tag{34}$$

where $\bar{f}$ is a reference image. When $\bar{f}$ is a constant array, then the cross-entropy becomes the negative Shannon entropy considered, for instance, in [44]. If both f and $\bar{f}$ satisfy the constraint (4), then a possible choice for the functions U_1, V_1 is

$$U_1(n) = -\ln \frac{f(n)}{c} \ , \quad V_1(n) = -\ln \frac{\bar{f}(n)}{c} \ , \tag{35}$$

where c is the flux constant defined in (4). We remark that, since the background is taken into account by the algorithms, f can be zero in some pixels; for this reason in the computation of the gradient we add a small quantity to the values of f. We also remark that when $\bar{f}$ is a constant, e.g. c/N^2, then $V_1(n) = 2\ln N$.

- **Hypersurface (HS) regularization** [20]

$$J_1(f) = \sum_n \sqrt{\delta^2 + D^2(n)} \ , \quad \delta > 0 \ , \tag{36}$$

for which (5) holds by setting

$$U_1(n) = \frac{f(n_{1+}) + f(n_{2+})}{\sqrt{\delta^2 + D^2(n)}} + \frac{f(n_{1-})}{\sqrt{\delta^2 + D^2(n_{1-})}} + \frac{f(n_{2-})}{\sqrt{\delta^2 + D^2(n_{2-})}} \ ,$$

$$V_1(n) = \frac{2f(n)}{\sqrt{\delta^2 + D^2(n)}} + \frac{f(n)}{\sqrt{\delta^2 + D^2(n_{1-})}} + \frac{f(n)}{\sqrt{\delta^2 + D^2(n_{2-})}} \ .$$

The application of SGP to the case of HS regularization is already considered in [15] and [4] for a comparison of its accuracy with that of Total Variation (TV) regularization.

- **Markov random field (MRF) regularization** [27]

$$J_1(f) = \frac{1}{2} \sum_n \sum_{n' \in \mathcal{N}(n)} \sqrt{\delta^2 + \left(\frac{f(n) - f(n')}{\epsilon(n')}\right)^2} \, , \tag{37}$$

where $\delta > 0$, $\mathcal{N}(n)$ is a symmetric neighborhood made up of the eight first neighbors of n and $\epsilon(n')$ is equal to 1 for the horizontal and vertical neighbors and equal to $\sqrt{2}$ for the diagonal ones; thanks to the symmetry of $\mathcal{N}(n)$, Eq. (5) holds by setting

$$U_1(n) = \sum_{n' \in \mathcal{N}(n)} \frac{f(n')}{\epsilon(n')\sqrt{\delta^2 + \left(\frac{f(n')-f(n')}{\epsilon(n')}\right)^2}} \, ,$$

$$V_1(n) = \sum_{n' \in \mathcal{N}(n)} \frac{f(n)}{\epsilon(n')\sqrt{\delta^2 + \left(\frac{f(n)-f(n')}{\epsilon(n')}\right)^2}} \, .$$

- **MISTRAL regularization (MIST)** [38]

$$J_1(f) = \sum_n \left\{ |D(n)| - \delta \ln \left(1 + \frac{|D(n)|}{\delta}\right) \right\} \, , \quad \delta > 0 \, , \tag{38}$$

for which (5) holds by setting

$$U_1(n) = \frac{f(n_{1+}) + f(n_{2+})}{\delta + |D(n)|} + \frac{f(n_{1-})}{\delta + |D(n_{1-})|} + \frac{f(n_{2-})}{\delta + |D(n_{2-})|} \, ,$$

$$V_1(n) = \frac{2f(n)}{\delta + |D(n)|} + \frac{f(n)}{\delta + |D(n_{1-})|} + \frac{f(n)}{\delta + |D(n_{2-})|} \, .$$

References

1. Anconelli, B., Bertero, M., Boccacci, P., Carbillet, M., Lantéri, H.: Reduction of boundary effects in multiple image deconvolution with an application to LBT LINC-NIRVANA. Astron. Astrophys. **448**, 1217–1224 (2006)
2. Bardsley, J.M., Goldes, J.: Regularization parameter selection methods for ill-posed poisson maximum likelihood estimation. Inverse Probl. **25**, 095,005 (2009)
3. Barzilai, J., Borwein, J.M.: Two-point step size gradient methods. IMA J. Numer. Anal. **8**(1), 141–148 (1988)
4. Benfenati, A., Ruggiero, V.: Inexact bregman iteration with an application to poisson data reconstruction. Inverse Probl. **29**, 065,016 (2013)
5. Bertero, M., Boccacci, P.: Introduction to Inverse Problems in Imaging. IoP Publishing, Bristol (1998)
6. Bertero, M., Boccacci, P.: A simple method for the reduction of boundary effects in the Richardson-Lucy approach to image deconvolution. Astron. Astrophys. **437**, 369–374 (2005)
7. Bertero, M., Boccacci, P., Desiderà, G., Vicidomini, G.: Image deblurring with Poisson data: from cells to galaxies. Inverse Probl. **25**(12), 123,006 (2009)
8. Bertero, M., Boccacci, P., La Camera, A., Olivieri, C., Carbillet, M.: Imaging with LINC-NIRVANA, the Fizeau interferometer of the Large Binocular Telescope: state of the art and open problems. Inverse Probl. **27**(11), 113,011 (2011)
9. Bertero, M., Boccacci, P., Talenti, G., Zanella, R., Zanni, L.: A discrepancy principle for Poisson data. Inverse Probl. **26**(10), 10,500 (2010)

10. Bertsekas, D.: Nonlinear Programming. Athena Scientific, Belmont (1999)
11. Bonettini, S., Landi, G., Loli Piccolomini, E., Zanni, L.: Scaling techniques for gradient projection-type methods in astronomical image deblurring. Int. J. Comput. Math. **90**(1), 9–29 (2013)
12. Bonettini, S., Prato, M.: Nonnegative image reconstruction from sparse Fourier data: a new deconvolution algorithm. Inverse Probl. **26**(9), 095,001 (2010)
13. Bonettini, S., Prato, M.: New convergence results for the scaled gradient projection method. Inverse Probl. **31**(9), 095,008 (2015)
14. Bonettini, S., Ruggiero, V.: On the uniqueness of the solution of image reconstruction problems with Poisson data. In: Simos, T.E., Psihoyios, G., Tsitouras, C. (eds.), International Conference of Numerical Analysis and Applied Mathematics 2010, AIP Conference Proceedings, vol. 1281, pp. 1803–1806 (2010)
15. Bonettini, S., Ruggiero, V.: An alternating extragradient method for total variation-based image restoration. Inverse Problems **27**, 095,001 (2011)
16. Bonettini, S., Zanella, R., Zanni, L.: A scaled gradient projection method for constrained image deblurring. Inverse Probl. **25**(1), 015,002 (2009)
17. Byrne, C.L.: Iterative image reconstruction algorithms based on cross-entropy minimization. IEEE Trans. Image Proc. **2**(1), 96–103 (1993)
18. Carasso, A.S.: Linear and nonlinear image deblurring: a documented study. SIAM J. Numer. Anal. **36**(6), 1659–1689 (1999)
19. Carbillet, M., La Camera, A., Deguignet, J., Prato, M., Bertero, M., Aristidi, E., Boccacci, P.: Strehl-constrained reconstruction of post-adaptive optics data and the Software Package AIRY, v. 6.1. In: Marchetti, E., Close, L.M., Véran, J.P. (eds.), Adaptive Optics Systems IV. Proceedings of SPIE, vol. 9148, p. 91484U (2014)
20. Charbonnier, P., Blanc-Féraud, L., Aubert, G., Barlaud, M.: Deterministic edge-preserving regularization in computed imaging. IEEE T. Image Process. **6**, 298–311 (1997)
21. Conrad, A., de Kleer, K., Leisenring, J., La Camera, A., Arcidiacono, C., Bertero, M., Boccacci, P., Defrère, D., de Pater, I., Hinz, P., Hofmann, K.H., Kürster, M., Rathbun, J., Schertl, D., Skemer, A., Skrutskie, M., Spencer, J., Veillet, C., Weigelt, G., Woodward, C.E.: Spatially Resolved M-Band Emission from Io's Loki Patera-Fizeau Imaging at the 22.8m LBT. Astron. J. **149**(5), 1–9 (2015)
22. Cornelio, A., Porta, F., Prato, M.: A convergent least-squares regularized blind deconvolution approach. Appl. Math. Comput. **259**(12), 173–186 (2015)
23. Correia, S., Carbillet, M., Boccacci, P., Bertero, M., Fini, L.: Restoration of interferometric images: I. the software package AIRY. Astron. Astrophys. **387**, 733–743 (2002)
24. Csiszár, I.: Why least squares and maximum entropy? An axiomatic approach to inference for linear inverse problems. Ann. Stat. **19**(4), 2032–2066 (1991)
25. De Mol, C., Defrise, M.: Inverse imaging with mixed penalties. In: Proceedings of the International Symposium on Electromagnetic Theory, pp. 798–800. Pisa, Italy (2004)
26. Engl, H.W., Hanke, M., Neubauer, A.: Regularization of Inverse Problems. Kluver Academic Publishers, Dordrecht (1996)
27. Geman, S., Geman, D.: Stochastic relaxation, Gibbs distributions and the Bayesian restoration of images. IEEE Trans. Pattern Anal. Mach. Intell. **6**(6), 721–741 (1984)
28. Giovannelli, J.F., Coulais, A.: Positive deconvolution for superimposed extended source and point sources. Astron. Astrophys. **439**, 401–412 (2005)
29. Herbst, T., Ragazzoni, R., Andersen, D., Boehnhardt, H., Bizenberger, P., Eckart, A., Gaessler, W., Rix, H.W., Rohloff, R.R., Salinari, P., Soci, R., Straubmeier, C., Xu, W.: LINC-NIRVANA: a fizeau beam combiner for the large binocular telescope. In: W.A. Traub (ed.), Interferometry for Optical Astronomy II. Proceedings of SPIE, vol. 4838, pp. 456–465 (2003)
30. Herbst, T.M., Santhakumari, K.K.R., Klettke, M., Arcidiacono, C., Bergomi, M., Bertram, T., Berwein, J., Bizenberger, P., Briegel, F., Farinato, J., Marafatto, L., Mathar, R., McGurk, R., Ragazzoni, R., Viotto, V.: Commissioning multi-conjugate adaptive optics with LINC-NIRVANA on LBT. In: Adaptive Optics Systems VI. Society of Photo-Optical Instrumentation Engineers (SPIE) Conference Series, vol. 10703, p. 107030B (2018). https://doi.org/10.1117/12.2313421

31. Hill, J.M., Green, R.F., Slagle, J.H.: The large binocular telescope. In: Ground-based and Airborne Telescopes. Proceedings of SPIE, vol. 6267, p. 62670Y (2006)
32. Hinz, P., Bippert-Plymate, T., Breuninger, A., Connors, T., Duffy, B., Esposito, S., Hoffmann, W., Kim, J., Kraus, J., McMahon, T., Montoya, M., Nash, R., Durney, O., Solheid, E., Tozzi, A., Vaitheeswaran, V.: Status of the LBT interferometer. In: Optical and Infrared Interferometry. Proceedings of SPIE, vol. 7013, p. 701328 (2008)
33. Hom, E.F.Y., Marchis, F., Lee, T.K., Haase, S., Agard, D.A., Sedat, J.W.: Aida: an adaptive image deconvolution algorithm with application to multi-frame and three-dimensional data. J. Opt. Soc. Am. **A-24**, 1580–1600 (1994)
34. Jefferies, S.M., Christou, J.C.: Restoration of astronomical images by iterative blind deconvolution. Astron. J. **415**, 609–862 (1993)
35. La Camera, A., Antonucci, S., Bertero, M., Boccacci, P., Lorenzetti, D., Nisini, B.: Image reconstruction for observations with a high dynamic range: LINC-NIRVANA simulations of a stellar jet. In: Delplancke, F., Rajagopal, F.J.K., Malbet, F. (eds.) Optical and Infrared Interferometry III. Proceedings of SPIE, vol. 8455, p. 84553D (2012)
36. Lantéri, H., Roche, M., Aime, C.: Penalized maximum likelihood image restoration with positivity constraints: multiplicative algorithms. Inverse Probl. **18**(5), 1397–1419 (2002)
37. Leisenring, J.M., Hinz, P.M., Skrutskie, M.F., Skemer, A., Woodward, C.E., Veillet, C., Arcidiacono, C., Bailey, V., Bertero, M., Boccacci, P., Conrad, A., de Kleer, K., de Pater, I., Defrère, D., Hill, J., Hofmann, K.H., Kaltenegger, L., La Camera, A., Nelson, M.J., Schertl, D., Spencer, J., Weigelt, G., Wilson, J.C.: Fizeau interferometric imaging of Io volcanism with LBTI/LMIRcam. In: Optical and Infrared Interferometry IV. Proceedings of SPIE, vol. 9146, p. 91462S (2014)
38. Mugnier, L.M., Fusco, T., Conan, J.M.: Mistral: a myopic edge-preserving image restoration method, with application to astronomical adaptive-optics-corrected long-exposure images. J. Opt. Soc. Am. **A-21**(4), 1841–1854 (2004)
39. Porta, F., Prato, M., Zanni, L.: A new steplength selection for scaled gradient methods with application to image deblurring. J. Sci. Comput. **65**(3), 895–919 (2015)
40. Porta, F., Zanella, R., Zanghirati, G., Zanni, L.: Limited-memory scaled gradient projection methods for real-time image deconvolution in microscopy. Commun. Nonlinear Sci. Numer. Simul. **21**(1–3), 112–127 (2015)
41. Prato, M., Cavicchioli, R., Zanni, L., Boccacci, P., Bertero, M.: Efficient deconvolution methods for astronomical imaging: algorithms and IDL-GPU codes. Astron. Astrophys. **539**, A133 (2012)
42. Prato, M., La Camera, A., Bonettini, S., Bertero, M.: A convergent blind deconvolution method for post-adaptive-optics astronomical imaging. Inverse Probl. **29**(6), 065,017 (2013)
43. Prato, M., La Camera, A., Bonettini, S., Rebegoldi, S., Bertero, M., Boccacci, P.: A blind deconvolution method for ground based telescopes and fizeau interferometers. New Astron. **40**, 1–13 (2015)
44. Skilling, J., Bryan, R.K.: Maximum entropy image reconstruction: general algorithm. Mon. Not. R. Astr. Soc. **211**, 111–124 (1984)
45. Skrutskie, M.F., Jones, T., Hinz, P., Garnavich, P., Wilson, J., Nelson, M., Solheid, E., Durney, O., Hoffmann, W.F., Vaitheeswaran, V., McMahon, T., Leisenring, J., Wong, A., Garnavic, P., Wilson, J., Nelson, M., Sollheid, E., Durney, O., Hoffmann, W.F., Vaitheeswaran, V., McMahon, T., Leisenring, J., Wong, A.: The Large Binocular Telescope mid-infrared camera (LMIRcam): final design and status. In: McLean, I.S., Ramsay, S.K., Takami, H. (eds.) Ground-based and Airborne Instrumentation for Astronomy III. Proceedings of SPIE, vol. 7735, p. 77353H (2010). https://doi.org/10.1117/12.857615. http://proceedings.spiedigitallibrary.org/proceeding.aspx?articleid=750817
46. Snyder, D.L., Hammoud, A.M., White, R.L.: Image recovery from data acquired with a charge-coupled-device camera. J. Opt. Soc. Am. A **10**(5), 1014–1023 (1994)
47. Snyder, D.L., Helstrom, C.W., Lanterman, A.D., Faisal, M., White, R.L.: Compensation for readout noise in CCD images. J. Opt. Soc. Am. A **12**(2), 272–283 (1995)

48. Titterington, D.: On the iterative image space reconstruction algorithm for ECT. IEEE Trans. Med. Imaging **6**(1), 52–56 (1987)
49. Vio, R., Bardsley, J., Wamsteher, W.: Least-squares methods with poissonian noise: analysis and comparison with the richardson-lucy algorithm. Astron. Astrophys. **436**(2), 741–755 (2005)
50. Zanella, R., Boccacci, P., Zanni, L., Bertero, M.: Efficient gradient projection methods for edge-preserving removal of Poisson noise. Inverse Probl. **25**(4), 045,010 (2009)

On the Segmentation of Astronomical Images via Level-Set Methods

Silvia Tozza and Maurizio Falcone

Abstract Astronomical images are of crucial importance for astronomers since they contain a lot of information about celestial bodies that can not be directly accessible. Most of the information available for the analysis of these objects starts with sky explorations via telescopes and satellites. Unfortunately, the quality of astronomical images is usually very low with respect to other real images and this is due to technical and physical features related to their acquisition process. This increases the percentage of noise and makes more difficult to use directly standard segmentation methods on the original image. In this work we will describe how to process astronomical images in two steps: in the first step we improve the image quality by a rescaling of light intensity whereas in the second step we apply level-set methods to identify the objects. Several experiments will show the effectiveness of this procedure and the results obtained via various discretization techniques for level-set equations.

Keywords Image segmentation · Level-set methods · Semi-lagrangian schemes · Finite difference schemes · Astronomical images

1 Introduction

Astronomical images are acquired by appropriate sensors, called CCDs (*Charge-Coupled Devices*), that are able to generate an electric charge at each pixel. This

The authors are members of the INdAM Research group GNCS.

S. Tozza (✉)
Istituto Nazionale di Alta Matematica, U.O. Dipartimento di Matematica, "Sapienza" Università di Roma,
P. le Aldo Moro, 5, 00185 Rome, Italy
e-mail: tozza@mat.uniroma1.it

M. Falcone
Dipartimento di Matematica, "Sapienza" Università di Roma, P. le Aldo Moro, 5, 00185 Rome, Italy
e-mail: falcone@mat.uniroma1.it

charge is directly proportional to the electromagnetic radiation that affects the pixel and is the measure corresponding to the "brightness" of real optical images. A typical feature of astronomical images is that they suffer from various types of noise which make difficult to analyze them. Their noise percentage is usually much higher than that of standard optical images since the value at every pixel does not correspond to the flow of photons emitted from the light source, i.e. the real signal, but is modified by the disturbances in the acquisition process. Let us recall the most important disturbances:

- the noise related to the signal, modeled by a Poisson distribution with standard deviation $\sqrt{n_e}$, which is directly proportional to the flux emitted by the source
- the light coming from other celestial bodies and from the sky, i.e. the spurious light collected by the telescope (the so-called *sky background*)
- the thermal noise, caused by overheating of the CCD sensors, which leads to an increase of the thermal agitation and the generation of additional conduction electrons;
- the *readout noise*, caused by the electronic components of the CCD and due to the discrete nature of the signal.

The amount of noise present in the image is expressed mathematically in terms of SNR (*Signal to Noise Ratio*), defined as the ratio between the power of the represented signal and that of the estimated noise, considering all the components previously listed. Larger values for this ratio correspond to images of better quality. The original image can not be used for an accurate scientific analysis of the data as we will see in the following sections. For that reason, a series of preprocessing steps are performed to reduce the noise and improve the image quality. It has been shown that, by increasing the exposure time of the sensors to light, the ratio between signal and noise can be greatly increased. This improvement is directly proportional to $\sqrt{t_{exp}}$ but an exposure time that is too long can lead to a saturation of the pixels so this procedure has to be carefully implemented. Furthermore, noise reduction operations are performed on each image. Typical operations include masking the defective pixels, subtracting the estimated value for the sky background and calibrating the image, but one can also apply a standard (linear or nonlinear) filter as we will do in our experiments. After these operations the value of the flow, with its relative uncertainty, and the astronomical coordinates associated to each pixel are redefined. Among the many other precautions that can be used, we emphasize that the most recent astronomical instruments use cooling devices for the CCD sensors which allow to reduce the readout noise. Despite the operations of calibration and noise reduction and the wide variety of techniques that has been adopted, noise remains one of the main components of the astronomical images. Due to the above steps in the acquisition, the range for the admissible values for the astronomical images is really different from the range of other kinds of images, e.g. it is common to have negative values at some pixels after the subtraction of the sky background. A final difference with respect to classical images is the format currently used to store astronomical images. The most common format is FITS (*Flexible Image Transport System*, [19]), introduced by the International Astronomical Union FITS Working Group (IAUFWG) in 1981 and up-dated in 2016 to its fourth version. The introduction of a new format is due to the need of save different information related to the images generated through CCD sensors, such as the angular coordinates of the portion of sky observed or the zero-point magnitude of the sensor used. This makes necessary to establish a common format, through which all the astronomers can interpret the data in the same way. The format has been developed so that all files, even the oldest ones, can be read from every machine, structuring files as a sequence of logical data.

To set this paper into perspective, let us mention some related contributions in the literature. The problem of deblurring astronomic images produced by telescopes is a classical and difficult problem in the astronomical community [12, 25, 30]. A novel technique to reduce the distortion caused by the ground-level turbolence of the atmosphere has been recently proposed in [17]. A similar goal has motivated the development of a high-resolution speckle imaging technique presented in [14]. As far as segmentation models is concerned, we mention that a modified version of the Chan-Vese model [7] has been proposed and analyzed in [13], some results obtained by a high-order splitting scheme are also presented there. It is interesting to note that this is a region based method with a level-set representation that can be applied to multispectral images.

In this paper we propose a strategy to analyze and segment astronomical images via the level-set method introduced in [21]. Although the segmentation problem has been investigated by many authors (see e.g the monographies [8, 20, 28] and the references therein) and several successful applications have been reported in many areas, level-set techniques are still not very popular in the astronomers community. Most probably this is due to the above mentioned features of astronomical images that make a direct application of these methods fail or give inaccurate results. Here we propose a coupling between an appropriate rescaling technique and a standard level-set methods to improve the global accuracy of the segmentation and increase the number of celestial bodies that can be extracted from a single image. We also add a filtering step to reduce the noise before segmenting. Hopefully, this will help astronomers in their sky investigations.

The paper is organized as follows: In Sect. 2, we propose new different rescaling transformations, adopted as the first two steps of our algorithm to improve the results of the segmentation of astronomical images. We briefly recall in Sect. 3 the first and second order equations related to level-set methods and the corresponding finite difference and semi-Lagrangian schemes that we used for our numerical experiments, at the beginning of this section we give some hints on the filtering step. Finally, in Sect. 4, we present our complete Rescaling Segmentation Algorithm (RSA) and we discuss in detail our numerical tests on simulated and real astronomical images. We conclude with Sect. 5 where we summarize our final remarks and future perspectives.

2 Efficient Rescaling of Astronomical Images

Let us start describing the first step of the procedure we adopted to segment astronomical images. It is useful to read astronomical images saved in the FITS format in MATLAB, thanks to the command *fitsread* and transform them in the matrix format that is common in image processing. The matrix I_0 returned as output from the function *fitsread* can take negative values due to the preprocessing techniques of calibration and reduction of noise applied to the images provided by the CCD sensors (e.g. procedures as the calibration or the subtraction operation of the estimated sky background).

We need to rescale the image values, defined on a rectangular domain $\overline{\Omega}$, with $\Omega \subset \mathbb{R}^2$, in order to obtain real values in $[0, 1]$. Starting from the matrix I_0, this is done defining

$$\widetilde{I_0} = \frac{I_0(x, y) - m_0}{M_0 - m_0}, \tag{1}$$

where

$$m_0 := \min_{(x,y) \in \overline{\Omega}} I_0(x, y), \quad M_0 := \max_{(x,y) \in \overline{\Omega}} I_0(x, y).$$

The image $\widetilde{I_0}$ obtained is still not ready for the segmentation since, in most cases, is very dark and only few celestial bodies will be visible to the naked eye. For that reason, we choose to transform the image, rescaling the values of the pixels by means of an appropriate function that we will construct in the sequel.

2.1 Elevation to Power or Logarithmic Rescaling

We look for a rescaling function $r : [0, 1] \rightarrow \mathbb{R}$ for the gray levels. Since these values for the image $\widetilde{I_0}$ obtained by (1) are in the range $[0, 1]$, the function r must satisfy the following conditions:

A1. $r([0, 1]) \subseteq [0, 1]$
A2. $r(0) = 0, r(1) = 1$
A3. r strictly increasing.

In other words, the rescaling transformation must keep the minimum and maximum brightness points of the image unaltered and rescale the intermediate values, without changing their ordering. Since the image is very dark, we also want the transformation to amplify the brightness values. In mathematical terms, we require r to satisfy the additional condition

A4. $r(x) > x, \quad \forall x \in [0, 1]$.

Clearly, several functions can satisfy the above four properties. A simple choice is given by

$$r_1(x) := x^\alpha, \tag{2}$$

with $\alpha \in (0, 1)$ a fixed parameter.

Another function can be obtained by a logarithmic transformation of the form

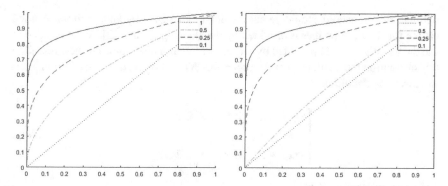

Fig. 1 Performance of the functions r_1 and r_2, by varying the parameter $\alpha \in (0, 1]$

$$r_2(x) := \left[\frac{\ln(x + 1)}{\ln 2} \right]^{\alpha}, \tag{3}$$

with $\alpha \in (0, 1)$. Both functions converge pointwise to the identity for α going to 1, whereas for α going to 0 they converge pointwise to the function

$$\widetilde{r_1}(x) := \begin{cases} 0 & \text{if } x = 0, \\ 1 & \text{if } x \in (0, 1]. \end{cases} \tag{4}$$

The latter transforms every brightness value, with the exception of the null one, assigning it the value 1. After the rescaling, the brightness increases as α decreases. The behavior of the two functions for different values of α is visible in Fig. 1.

In the numerical tests presented in Sect. 4, we will only show the results obtained with the function r_1, since for the same α, r_2 gives almost identical results.

2.2 Rescaling with a Threshold

From our experiments on astronomical images (see Sect. 4) we have observed that the proposed transformations r_1 and r_2 can be improved. As we said, astronomical images are affected by a strong noise component and the rescaling has a significant effect also on high brightness values due to the noise component. When these values are rescaled, they result too high so the global effect is an amplification of the tone differences with respect to the pixels closer to the real tone of the background. This amplification can make the segmentation method fail, identifying artificial objects that are not present in the real image. To avoid this undesired effect the rescaling should distinguish the pixels of celestial bodies from those of the background: the values of the former must be amplified, while the others must be attenuated. A natural idea is to introduce a threshold to determine the gray tones of the objects and, to be optimal, this threshold should be automatically identified by an algorithm. A good choice for standard images is provided by the Otsu algorithm [22], so we decided to use the value $\tau \in [0, 1)$ provided by this algorithm as a threshold. The new rescaling transformation must still respect the properties A1–A3 of Sect. 2.1, in addition it has to satisfy the condition

$$\begin{cases} r(x) < x, & \text{if } 0 < x < \tau, \\ r(\tau) = \tau, \\ r(x) > x, & \text{if } \tau < x < 1, \end{cases} \tag{A4$_\tau$}$$

and to be continuous at $x = \tau$.

A rescaling function of this type can be obtained by considering the applications x^β and $x^{1/\beta}$, with $\beta \in \mathbb{N} \setminus \{0\}$, respectively in the two subsets $[0, \tau]$ and $(\tau, 1]$. These functions have to be appropriately translated and expanded to respect all the conditions. In this way, we obtain the function

$$r_3(x) := \begin{cases} \dfrac{x^\beta}{\tau^{\beta-1}}, & \text{if } 0 \le x < \tau, \\ \dfrac{(x - \tau)^{1/\beta}}{(1 - \tau)^{1/\beta-1}} + \tau, & \text{if } \tau \le x \le 1. \end{cases} \tag{5}$$

The behavior of r_3 varying $\beta \in \mathbb{N} \setminus \{0\}$ is reported in Fig. 2. This function satisfies the properties listed before, converges pointwise to the identity function for β tending to 1 and to the following function

$$\tilde{r}_3(x) := \begin{cases} 0, & \text{if } 0 \le x < \tau, \\ \tau, & \text{if } x = \tau, \\ 1, & \text{if } \tau < x \le 1, \end{cases} \tag{6}$$

for β tending to $+\infty$.

3 Segmentation via Level-Set Methods

As we said in the introduction, we follow the level-set (LS) approach to segmentation problems obtained by the rescaling procedure described in the previous section. For readers convenience let us briefly describe the main features of this approach. The level-set method has been introduced by Osher and Sethian [21, 27] and since then it has been widely used in many applications, e.g. fronts propagation, computer vision, computational fluids dynamics (see [20, 28] for several interesting examples). Its popularity is due to the simplicity in the implementation and its capability to follow topological changes (splitting, merging) in time. Typical examples are when a planar curve (or a multidimensional surface) splits into many parts or when several evolving

Fig. 2 Behavior of the function r_3 by varying the parameter $\beta \in \mathbb{N} \setminus \{0\}$

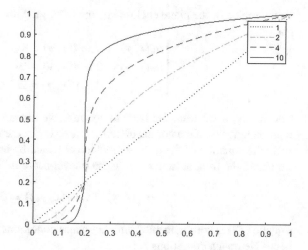

curves (or surfaces) merge into a single one. This is the main reason for its popularity also in the image processing community. For the segmentation problem the idea is to define a normal vector field bringing an initial curve (e.g. a circle) onto the object boundaries in an image.

Let us consider an image $\widetilde{I_0} : \overline{\Omega} \to [0, 1]$, with $\Omega \subset \mathbb{R}^2$ an open rectangular domain. Let us fix an initial curve $\gamma_0 \subset \overline{\Omega}$. We want to track its evolution according to the normal velocity and we define it so that it goes to zero (and therefore the front stops) at the edges of the object to be identified. The methods based on this approach can be divided into two subclasses. In the methods belonging to the first class, the speed depends on the gradient of the image $\widetilde{I_0}$ at each point $(x, y) \in \Omega$, since the gradient provides a measure of the gray-level variation in the image and therefore it identifies the presence of edges. The second class of methods, introduced by Chan and Vese [7], is inspired by a variational segmentation technique proposed by Mumford and Shah [18] and is based on the minimization of a functional which allows to partition the image in regions where there is a small variation of gray levels. Looking more in details the first class, we have to solve an evolutive Hamilton–Jacobi equation

$$\begin{cases} u_t(x, y, t) + c(x, y, t)|\nabla u(x, y, t)| = 0, & \forall (x, y, t) \in \Omega \times (0, T], \\ u(x, y, 0) = u_0(x, y), & \forall (x, y) \in \overline{\Omega}, \end{cases} \tag{7}$$

with $u(\cdot, \cdot, t)$ and u_0 the representation function of the front at time t and at the initial time, respectively, and c is the velocity function. Depending on the definition for c (that in general may depend on x, t and the curvature), Eq. (7) will be a first or second order equation. Several explicit definitions of c will be reported in Sect. 3.2. In order to segment a given image $\widetilde{I_0}$, we choose the initial front $\gamma_0 \subset \overline{\Omega}$ and its representation function $u_0 : \overline{\Omega} \to \mathbb{R}$. In particular, if we want to approximate the edges of the object with a curve that expands from within, we choose u_0 in such a way that, denoted by

ω_0 the region of the plane enclosed by the front γ_0, with $\gamma_0 = \partial \omega_0$ and ω_0 open, we put

$$\begin{cases} u_0(x, y) < 0, & \forall (x, y) \in \omega_0, \\ u_0(x, y) = 0, & \forall (x, y) \in \gamma_0, \\ u_0(x, y) > 0, & \forall (x, y) \in \overline{\Omega} \setminus \overline{\omega_0}. \end{cases} \qquad (8)$$

Conversely, if we want the front to contract, we can reverse the sign of the initial representative or of the normal direction. Next, we fix the velocity of the front and we solve the equation of the level-set method obtained by it: denoted by u its solution, we obtain the front at time $t > 0$ as the 0-level-set of $u(\cdot, \cdot, t)$, that is

$$\gamma_t = \left\{ (x, y) \in \overline{\Omega} \mid u(x, y, t) = 0 \right\}. \qquad (9)$$

Equation (7) is complemented with boundary conditions. We chose to use homogeneous Neumann conditions

$$\frac{\partial u}{\partial \eta}(x, y, t) = 0, \quad \forall (x, y, t) \in \partial \Omega \times (0, T]. \qquad (10)$$

The choice of the final time T to which numerically solve the Eq. (7) will have to be carried out through a stopping criterion, which detects when the front is near equilibrium, through the verification of a condition. In this paper, we will adopt the following criterion: First, at each iteration we identify the grid nodes near the front with respect to a fixed tolerance denoted by ε_F. More precisely, since the front at time t_n is the 0-level curve of the representation function, we define the approximate front by means of $\mathbf{V}^n := \{v_{i,j}^n\}$, where $v_{i,j}^n$ is the value computed on the grid node (x_i, y_j) at time n. Our numerical front is given by

$$F^n \equiv \{(x_i, y_j) : |v_{i,j}^n| \leq \varepsilon_F \}. \qquad (11)$$

Let us denote by $\mathscr{F}$ the set of indexes of the nodes that respect this condition and with $\mathbf{V}^{n,F}$ the vector formed by the elements of $\mathbf{V}^n$ corresponding to them. Hence, we fix an additional tolerance ε: the stopping condition of the numerical scheme will be

$$\left\| \mathbf{V}^{n+1,F} - \mathbf{V}^{n,F} \right\|_1 \leq \varepsilon, \qquad (12)$$

with the norm $\| \cdot \|_1$ defined by

$$\left\| \mathbf{V}^{n+1,F} - \mathbf{V}^{n,F} \right\|_1 := \Delta x^2 \sum_{(i,j) \in \mathscr{F}} \left| v_{i,j}^{n+1} - v_{i,j}^n \right|. \qquad (13)$$

In other words, we proceed to solve the scheme up to the $(n + 1)$-th iteration when the representation has reached equilibrium with a tolerance ε at all the nodes belonging to F^n.

3.1 The Filtering Pre-processing Step

Let us analyze the first class of active contours methods. Since the edges of objects are, in most cases, identified by large variations of gray tones in their neighborhood, we can define the velocity of the front as a function of the gradient of the function $\widetilde{I_0}$ that models the image. However, $\widetilde{I_0}$ is a noisy image so in order to define its gradient it is useful to add a filtering step on it. We did it in two different ways: by applying a Gaussian filter, i.e. solving the *heat equation with homogeneous Neumann conditions*

$$\begin{cases} I_t(x, y, t) = \Delta I(x, y, t), & \forall (x, y, t) \in \Omega \times (0, T_C], \\ \dfrac{\partial I}{\partial \eta}(x, y, t) = 0, & \forall (x, y, t) \in \partial\Omega \times (0, T_C], \\ I(x, y, 0) = \widetilde{I_0}(x, y), & \forall (x, y) \in \overline{\Omega}, \end{cases} \tag{14}$$

which has a diffusive effect on the initial datum $\widetilde{I_0}$, for a small fixed time $T_c > 0$ (in the numerical tests, it is of the order of 10^{-3} or 10^{-4}). Numerically, we solve (14) by the standard centered finite difference scheme

$$I_{i,j}^{n+1} = I_{i,j}^n + \widetilde{\Delta t} \left[\frac{I_{i+1,j}^n + I_{i,j+1}^n - 4I_{i,j}^n + I_{i-1,j}^n + I_{i,j-1}^n}{\Delta x^2} \right], \tag{15}$$

forward in time, with time step $\widetilde{\Delta t} > 0$ and space steps $\Delta x = \Delta y$. In (15) $I_{i,j}^n$ denotes as usual the approximation of the gray level of the image at the pixel of coordinate (i, j) and at time t_n, whereas $I_{i,j}^0 := \widetilde{I_0}(x_i, y_j)$ for every $(i, j) \in \mathscr{I}$, set of indexes. The required CFL condition for this numerical scheme is $\widetilde{\Delta t} \leq \Delta x^2/4$.

The consequence of applying the Gaussian filter is an edge blurring due to isotropic diffusion. Choosing large values of $|\nabla I|$ as an indicator of the edge points of the image, we would like to stop the diffusion at the edges, we pass from an isotropic to an anisotropic diffusion, i.e.

$$I_t = div(\nabla I) \text{ is replaced by } I_t = div(f(|\nabla I|)\nabla I). \tag{16}$$

This is the idea behind the Perona–Malik model [24] described by (16) and complemented by suitable boundary conditions (e.g. homogeneous Neumann boundary conditions), the initial condition is the original image. The anisotropic diffusion is driven by f and two typical choices for the diffusion coefficient are:

$$f_1(|\nabla I|) = \frac{1}{1 + \left(\frac{|\nabla I|}{\mu}\right)^2} \tag{17}$$

$$f_2(|\nabla I|) = \exp\left(-\left(\frac{|\nabla I|}{\mu}\right)^2\right) \tag{18}$$

where μ is the gradient magnitude threshold parameter. In our numerical simulations, we will use the function f_2. Let us denote by $\widetilde{I}_{filt}$ the solution of the problem (14) or (16), filtered version of the image $\widetilde{I}_0$.

3.2 Edge-Detector Functions

We want the velocity c of the front to vanish near the edges so we introduce a function g of $|\nabla \widetilde{I}_{filt}|$, called *edge-detector*, that has to satisfy the following conditions:

$$g : [0, +\infty) \to [0, +\infty) \text{ is decreasing and } \lim_{z \to +\infty} g(z) = 0 . \qquad (19)$$

In this way $g(|\nabla \widetilde{I}_{filt}(x, y)|)$ will tend to 0 approaching the points (x, y) near the edges to be identified, since at the edges we typically have very high values of $|\nabla \widetilde{I}_{filt}|$. Higher values of g will correspond to points where $|\nabla \widetilde{I}_{filt}| \approx 0$, i.e. to the regions where the gray tones of the image are approximately constant. Two possible choices for the edge-detector function are the following:

$$g_1(z) := \frac{1}{1 + z^p} , \quad \forall z \in [0, +\infty) , p \geq 1, \qquad (20)$$

proposed in [6] with $p = 2$, and in [16] with $p = 1$, and

$$g_2(z) := 1 - \frac{z - m}{M - m} , \quad \forall z \in [0, +\infty) , \qquad (21)$$

where

$$m := \min_{(x,y) \in \Omega} |\nabla \widetilde{I}_{filt}(x, y)| , \quad M := \max_{(x,y) \in \Omega} |\nabla \widetilde{I}_{filt}(x, y)|$$

defined in [16]. Practically, the values of $g_1(|\nabla \widetilde{I}_{filt}(x, y)|)$ vary between $(1 + M)^{-1}$ and $(1 + m)^{-1}$, whereas the values of $g_2(|\nabla \widetilde{I}_{filt}(x, y)|)$ are between 0 (for $|\nabla \widetilde{I}_{filt}| = M$) and 1 (for $|\nabla \widetilde{I}_{filt}| = m$).

Let us discuss some typical choices for the velocity c. A simple choice is to make c dependent just on the point

$$c(x, y, t) := g(x, y) , \quad \forall (x, y) \in \Omega . \qquad (22)$$

In this way, using the notation $g(x, y) := g(|\nabla \widetilde{I}_{filt}(x, y)|)$, the problem to solve becomes

$$\begin{cases} u_t(x, y, t) + g(x, y)|\nabla u(x, y, t)| = 0 , & \forall (x, y) \in \Omega , \forall t \in (0, T] , \\ \frac{\partial u}{\partial \eta}(x, y, t) = 0 , & \forall (x, y) \in \partial\Omega , \forall t \in (0, T] , \quad (23) \\ u(x, y, 0) = u_0(x, y) , & \forall (x, y) \in \overline{\Omega} , \end{cases}$$

with u_0 the representation function of the initial front. This is the isotropic case and the corresponding equation is a first-order Hamilton–Jacobi equation of eikonal type.

Another popular choice is to use a velocity that, at each point (x, y), depends on the geometric properties of the front, e.g. its curvature $k(x, y)$. This choice is more complicated since the velocity will also depend on u, hence on t. Following [1, 15], we consider a *curvature dependent velocity*

$$c(x, y, t) := g(x, y) (1 - vk(x, y)), \quad \forall (x, y) \in \Omega, \tag{24}$$

where $v > 0$ is a fixed parameter. The factor $g(x, y)$ causes that the front stops near the edges. The parameter v (typically less than 1) weighs the speed dependency on the curvature. Since the curvature is given by

$$k(x, y) = \text{div} \left(\frac{\nabla u(x, u, t)}{|\nabla u(x, y, t)|} \right), \quad \forall (x, y) \in \Omega, \tag{25}$$

the level-set corresponding equation is the second order Hamilton–Jacobi equation

$$\begin{cases} u_t(x, y, t) + g(x, y)|\nabla u(x, y, t)| = vg(x, y)\text{div}\left(\frac{\nabla u(x, u, t)}{|\nabla u(x, y, t)|}\right)|\nabla u(x, y, t)|, \\ \qquad\qquad\qquad\qquad\qquad\qquad\qquad \forall (x, y) \in \Omega, \, \forall t \in (0, T], \\ \frac{\partial u}{\partial \eta}(x, y, t) = 0, \qquad\qquad\qquad \forall (x, y) \in \partial\Omega, \, \forall t \in (0, T], \\ u(x, y, 0) = u_0(x, y), \qquad\qquad \forall (x, y) \in \overline{\Omega}, \end{cases} \tag{26}$$

with the same boundary conditions and initial datum as in (23). The term in the second member has a diffusive effect on the solution: consequently, this type of scheme can be useful for segmenting images characterized by noise. Note that, in practice, the function g is not necessarily equal to 0 at all points on the edges of the objects, even if it takes very small values. The stopping rule (12) allows to control the numerical scheme so that the evolution stops at time T whenever the velocity is below a given threshold.

In order to get a numerical solution of (23) and (26) in our tests we will use a finite difference scheme (FD) and a semi-Lagrangian scheme (SL), so we will be able to compare their results. Let us recall that the *FD schemes* for the two equations are, respectively,

$$v_{i,j}^{n+1} = v_{i,j}^n - \Delta t g_{i,j} \nabla^+, \tag{27}$$

and

$$\begin{cases} v_{i,j}^{n+1} = v_{i,j}^n - \Delta t g_{i,j} \nabla^+ + \frac{v}{4} g_{i,j} (v_{i+1,j}^n + v_{i,j+1}^n + v_{i-1,j}^n + v_{i,j-1}^n) \\ \qquad\qquad\qquad\qquad\qquad\qquad\qquad \text{if } |D_{i,j}^c[V^n]| \leq C\Delta x^s, \\ v_{i,j}^{n+1} = v_{i,j}^n - \Delta t g_{i,j} \nabla^+ + v\Delta t g_{i,j} \Lambda_{(i,j)}^n, \quad \text{if } |D_{i,j}^c[V^n]| > C\Delta x^s, \end{cases} \tag{28}$$

for each $(i, j) \in \mathscr{I}$ and $n \in \{0, 1, \ldots, N_T - 1\}$, where $\mathbf{V}^0 := \mathbf{U}(0)$, $g_{i,j} := g(\mathbf{x}_{i,j})$, with $\mathbf{x}_{i,j} := (x_i, y_j)$,

$$\nabla^+ := \big[\max\{D_{i,j}^{-x}[\mathbf{V}^n], 0\}^2 + \min\{D_{i,j}^{+x}[\mathbf{V}^n], 0\}^2 \qquad (29)$$
$$+ \max\{D_{i,j}^{-y}[\mathbf{V}^n], 0\}^2 + \min\{D_{i,j}^{+y}[\mathbf{V}^n], 0\}^2 \big]^{1/2}$$

and

$$\Lambda_{i,j}^n := \frac{1}{(D_{i,j}^{c,x}[\mathbf{V}^n]^2 + D_{i,j}^{c,y}[\mathbf{V}^n]^2)^{1/2}} \Big(D_{i,j}^{2,x}[\mathbf{V}^n] D_{i,j}^{c,y}[\mathbf{V}^n]^2 \qquad (30)$$
$$- 2D_{i,j}^{c,x}[\mathbf{V}^n] D_{i,j}^{c,y}[\mathbf{V}^n] D_{i,j}^{xy}[\mathbf{V}^n] + D_{i,j}^{2,y}[\mathbf{V}^n] D_{i,j}^{c,x}[\mathbf{V}^n]^2 \Big).$$

We refer the reader interested in the construction and the analysis of these schemes to [20, 28].

Let us also recall that the *SL schemes* are, respectively,

$$v_{i,j}^{n+1} = \min_{\mathbf{a} \in B(0,1)} \big\{ I[\mathbf{V}^n](\mathbf{x}_{i,j} - \Delta t g_{i,j} \mathbf{a}) \big\} \qquad (31)$$

and

$$\begin{cases} v_{i,j}^{n+1} = \min_{\mathbf{a} \in B(0,1)} \big\{ I[\mathbf{V}^n](\mathbf{x}_{i,j} - \Delta t g_{i,j} \mathbf{a}) \big\} + \frac{\nu}{4} g_{i,j} (v_{i+1,j}^n + v_{i,j+1}^n + v_{i-1,j}^n + v_{i,j-1}^n), \\ \qquad\qquad\qquad\qquad \text{if } |D_{i,j}^c[\mathbf{V}^n]| \leq C\Delta x^s, \\ v_{i,j}^{n+1} = \min_{\mathbf{a} \in B(0,1)} \big\{ I[\mathbf{V}^n](\mathbf{x}_{i,j} - \Delta t g_{i,j} \mathbf{a}) \big\} + \frac{\nu}{2} g_{i,j} \big[I[\mathbf{V}^n](\mathbf{x}_{i,j} + \sigma_{i,j}^n \sqrt{\Delta t}) \\ \quad + I[\mathbf{V}^n](\mathbf{x}_{i,j} - \sigma_{i,j}^n \sqrt{\Delta t}) \big], \qquad \text{if } |D_{i,j}^c[\mathbf{V}^n]| > C\Delta x^s, \end{cases} \qquad (32)$$

for each $(i, j) \in \mathscr{I}$ and $n \in \{0, 1, \ldots, N_T - 1\}$, with $\mathbf{V}^0 := \mathbf{U}(0)$ and

$$\sigma_{i,j}^n := \frac{\sqrt{2}}{|D_{i,j}^c[\mathbf{V}^n]|} \begin{pmatrix} D_{i,j}^{c,y}[\mathbf{V}^n] \\ -D_{i,j}^{c,x}[\mathbf{V}^n] \end{pmatrix}. \qquad (33)$$

The role of the threshold Δx^s for the first derivatives in (32) is explained in detail in [3] (see also [8] for the general theory of semi-Lagrangian schemes and [5] for other applications to image processing problems). For our purposes it is sufficient to note that this threshold is used to solve also the degenerate case without adding a regularization.

4 Numerical Tests

Let us start describing the complete segmentation algorithm that includes the rescaling preprocessing via the functions r_1, r_2 and r_3 illustrated in Sects. 2.1 and 2.2.

RESCALING SEGMENTATION ALGORITHM (RSA)

STEP 1: Apply to the original image I_0 the rescaling defined in (1) to get $\widetilde{I}_0$ which takes values in $[0, 1]$.

STEP 2: Choose one of the proposed rescaling functions r_i, $i \in \{1, 2, 3\}$, and set

$$\widetilde{I} := r_i(\widetilde{I}_0), \tag{34}$$

for each element of the matrix. In case we choose the function r_3, we first apply the thresholding method of Otsu to the image $\widetilde{I}_0$ in order to select the optimal threshold τ, and then we apply (34).

STEP 3: Filter the image $\widetilde{I}$ by few iterations of the linear filter given by the scheme (15), or applying the PM method (16) with f_2. This step produces $\widetilde{I}_{filt}$.

STEP 4: Apply one of the segmentation active contours methods to the image $\widetilde{I}_{filt}$ thus obtained in STEP 3.

We are now ready to present some numerical tests, using the RSA algorithm. Let us consider an $M \times N$ image and let us fix the discretization parameters as:

- $\Delta x = \Delta y = 0.1$ the space step of the uniform grid
- $\Delta t = \Delta x/4 = 0.025$, for the FD scheme approximating the first order problem
- $\Delta t = \Delta x^2 = 0.01$, for the FD scheme approximating the second order problem
- $\Delta t = \Delta x = 0.1$, for the SL schemes.

That structured grid has nodes located at the center of the pixels, with coordinates $((j-1)\Delta x, (i-1)\Delta x)$, for $j = 1, \ldots, M$ and $i = 1, \ldots, N$, and the rectangular domain is defined as

$$\Omega := \left[-\frac{\Delta x}{2}, a - \frac{\Delta x}{2} \right] \times \left[-\frac{\Delta x}{2}, b - \frac{\Delta x}{2} \right], \tag{35}$$

with $a := M\Delta x$ and $b := N\Delta x$. For each test, we will consider three cases:

- a segmentation of the original image, without rescaling (i.e. dropping STEP 2 of the algorithm, setting $\widetilde{I} = \widetilde{I}_0$)
- a segmentation using a rescaling of the image by r_1
- a segmentation using a rescaling of the image by r_3 and the optimal threshold computed by the Otsu's algorithm.

As we said in Sect. 2.1, we will omit the results obtained by rescaling via r_2 since the results are almost identical to that of r_1, with the same parameter α fixed.

For all the three cases listed above, before the segmentation step we filter the image by the linear or nonlinear filter described in Sect. 3.1. We will compare the performance of the four numerical schemes presented in Sect. 3.2. For comparison, we will also show the segmented image obtained by the software *SExtractor* [2], one of the most popular software in the astronomical community. In these images, each source is represented by the grey-level obtained as average of the pixels values that compose it. Since the images are too big, we will work on smaller images of size 300×300 pixels. Hence, we will have $a = b = 30$ and $N = M = 300$. For the

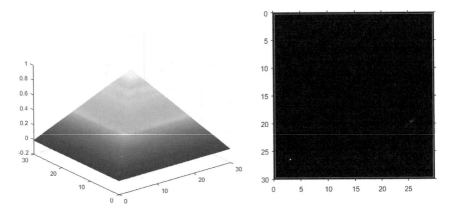

Fig. 3 Representation function u_0 and related front γ_0 at time $t = 0$

active contour method, we use a rectangular initial front γ_0, i.e. the external boundary of the image, described as the 0-level set of

$$u_0(x, y) := 1 - \left| \frac{x + y - 30}{29.6} \right| - \left| \frac{x - y}{29.6} \right|, \quad \forall (x, y) \in \Omega, \tag{36}$$

visible in Fig. 3, and we filter the image by applying 5 iterations of the scheme (15) or by 15 iterations of the PM method (16), with time step $\widetilde{\Delta t} = 10^{-4}$ unless otherwise stated. We need to fix two tolerances: $\varepsilon_F = \Delta x = 0.1$, for the identification of the nodes that approximate the front, and $\varepsilon = 10^{-3}$ for the stopping criterion. The maximum time when the scheme will not converge is set to $T_{max} = 50$.

For the edge-detector function g_1, the parameter $p \in \mathbb{N} \setminus \{0\}$ will be fixed according to the contrast in the image between objects and background. If objects are well defined, we set a value of p small, if the edges of the object have pixels with gray tones close to those of the background, the value of p has to be increased. The function g_2 defined in (21) in practice does not produce optimal results since it assumes null value only at points where the gradient of the image is maximum and this does not necessarily occur at all points belonging to the edges of the object. Therefore, as proposed in [4], we modify the function g_2, subtracting a constant $c_2 \in [0, 1]$ and then rescaling the values in $[0, 1]$. The function we use is the following

$$\widetilde{g_2}(z) := \frac{1}{1 - c_2} \max\{g_2(z) - c_2, 0\}. \tag{37}$$

Thanks to that definition, $\widetilde{g_2}(|\nabla \widetilde{I}_{filt}|)$ attains its maximum value equal to 1 for $|\nabla \widetilde{I}_{filt}| = m$, and null value when $|\nabla \widetilde{I}_{filt}|$ is greater than a fixed threshold, precisely $|\nabla \widetilde{I}_{filt}| \geq (1 - c_2)(M - m) + m$. In each test, we provide the values of the parameters involved, e.g. p for the function g_1, c_2 for the function $\widetilde{g_2}$, and ν for the dependence from the curvature in the second order schemes (28) and (32).

We acknowledge the contribution of L. Pecci to the implementation of the methods and to some of the tests presented here. Other numerical experiments are contained in [23].

Test 1: *f160.fits*

The first image, Fig. 4 on the left, is a cropping of a simulated, high resolution astronomical image provided by INAF (Istituto Nazionale di Astrofisica) and generated by reproducing data observed by the Hubble Space Telescope (HST). It depicts many stars, galaxies and other celestial bodies, as can be seen from the segmentation obtained with the software *SExtractor* in Fig. 4 on the right, although it is almost completely black in its original form. Our purpose is to apply a segmentation algorithm that traces as many sources as possible, possibly improving the results obtained by *SExtractor* thanks to the introduction of the proposed rescaling functions.

Test1: Without Rescaling

Let us start showing the results obtained by the four schemes considered, without using any rescale function. The original image *f160.fits* and the segmentation provided by the software *SExtractor* are shown in Fig. 4. Before applying the active contour schemes, we filter the original image *f160.fits* by using 5 iterations of the scheme (15) with a time step $\widetilde{\Delta t} = 10^{-4}$. All the active contour methods only identify the brightest celestial body or a few other objects. The results are shown in Figs. 5, 6, 7 and 8, the values of the parameters used are mentioned in the captions. We only show the results obtained by the schemes (27) and (31) with edge-stopping function $\widetilde{g}_2$ (Figs. 5, 6) and the second order schemes (28) and (32) with function g_1 (Figs. 7, 8). Even if we increase the values of the parameters p and c_2, we do not get better results. Due to the similarity, we decide to omit the results obtained by using the first order schemes (27) and (31) with edge-detector function g_1.

Note that by applying the PM method (16) to the original image, we do not get an improvement as shown in Fig. 9. It is important to note that the results without a rescaling preprocessing are really bad for all the schemes, even if we apply a nonlinear filtering algorithm.

Fig. 4 Test 1. From left to right: Image *f160.fits*, segmentation of the image provided by the software *SExtractor*

Fig. 5 Test 1 without rescaling: Position of the front at time $T = 15.85$ and segmented image, for the FD scheme (27) by using the edge-detector function $\widetilde{g_2}$, with $c_2 = 0.8$

Fig. 6 Test 1 without rescaling: Position of the front at time $T = 15$ and segmented image, for the SL scheme (31) by using the edge-detector function $\widetilde{g_2}$, with $c_2 = 0.8$

Fig. 7 Test 1 without rescaling: Position of the front at time $T = 16.61$ and segmented image, for the FD scheme (28), by using the edge-detector function g_1, $p = 5000$ and $\nu = 10^{-4}$

Fig. 8 Test 1 without rescaling: Position of the front at time $T = 15.4$ and segmented image, for the SL scheme (32), by using the edge-detector function g_1, $p = 5000$ and $\nu = 10^{-4}$

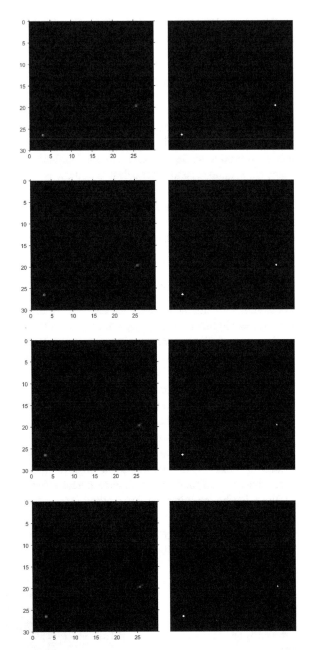

Fig. 9 Test 1 without rescaling: Position of the front at time $T = 15$ and segmented image, for the SL scheme (31) by using the edge-detector function $\widetilde{g_2}$, with $c_2 = 0.8$. Image filtered by 15 iterations of the PM method with f_2, for $\mu = 30$, and $\widetilde{\Delta t} = 10^{-4}$

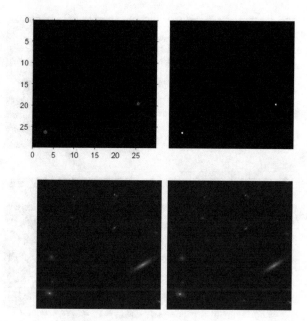

Fig. 10 Test 1. From left to right: Rescaling of the image *f160.fits* by using the function r_1, with $\alpha = 0.25$ and its filtered version obtained by 5 iterations of the scheme (15) with time step $\widetilde{\Delta t} = 10^{-4}$

Test 1: Rescaling by r_1

Let us test the algorithm rescaling the gray tones of the image before applying the segmentation methods. We use r_1 setting $\alpha = 0.25$ since for values of α too close to 1, the objects are not quite evident, whereas for smaller values the background tones are amplified excessively. The image obtained by the rescaling, shown in Fig. 10, is segmented via the four schemes listed before. Also in this case, we omit to show the results obtained by using the first order schemes with edge-detector function g_1 since this function fails even if we use a second order scheme, as we can see looking at Figs. 14, 15. In fact, g_1 does not identify the boundaries, even for large values of the parameter p, so that only very few objects are detected. Instead, the edge-detector function $\widetilde{g_2}$ (Figs. 11, 12) is able to identify a greater number of objects, even if the approximation of their contours is still non very accurate (for example for the larger galaxy, placed on the right of the image). Using the PM nonlinear filtering method after the rescaling process, we can note (comparing Figs. 13 and 12) that a better segmentation is obtained. In fact, more small objects are detected and visible in the final front and the associated segmented image: see e.g. two small red points in the central-bottom part of the final front in Fig. 13 not present in Fig. 12.

Test 1: Rescaling by r_3

Finally, we present the results obtained by r_3, this case seems to give the best results. The parameter chosen is $\beta = 8$, for which the boundaries of the objects appear more evident, with tones distant from those of the background (see Fig. 16). In this case,

Fig. 11 Test 1, rescaling by r_1: Position of the front at time $T = 18.075$ and segmented image, for the FD scheme (27) by using the edge-detector function $\widetilde{g_2}$, with constant $c_2 = 0.8$

Fig. 12 Test 1, rescaling by r_1: Position of the front at time $T = 18.1$ and segmented image, for the SL scheme (31) by using the edge-detector function $\widetilde{g_2}$, with $c_2 = 0.8$ filtered by the Gaussian filter

Fig. 13 Test 1, rescaling by r_1: Position of the front at time $T = 18.6$ and segmented image, for the SL scheme (31) by using the edge-detector function $\widetilde{g_2}$, with $c_2 = 0.8$. The rescaled image has been filtered by 15 iterations of the PM method with f_2, for $\mu = 30$, and $\widetilde{\Delta t} = 10^{-4}$

Fig. 14 Test 1, rescaling by r_1: Position of the front at time $T = 16.08$ and segmented image, for the FD scheme (28) by using the edge-detector function g_1, $p = 5000$ and $\nu = 10^{-6}$

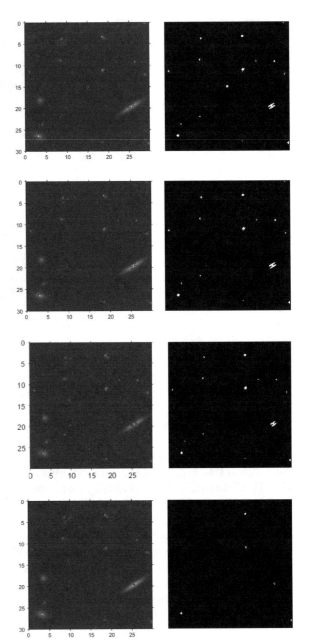

Fig. 15 Test 1, rescaling by r_1: Position of the front at time $T = 15.1$ and segmented image, for the SL scheme (32) by using the edge-detector function g_1 and $\nu = 10^{-6}$

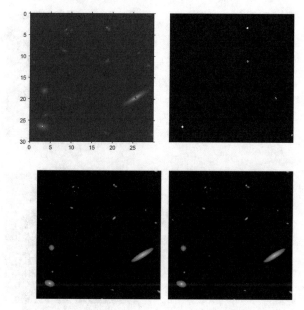

Fig. 16 Test 1. From left to right: Rescaling of the image *f160.fits* by using the function r_3, with $\beta = 8$, and its filtered version obtained by 5 iterations of the scheme (15) with time step $\widetilde{\Delta t} = 10^{-4}$

all the schemes seem to provide satisfactory results, even those based on the use of the edge-detector function g_1. Due to their similarity, also in this can we show only the two second order schemes with edge-detector function g_1. The resulting segmentations are illustrated in Figs. 17, 18, 19 and 20. These results show very well the improvements obtained by the rescaling r_3.

Test 2: *real.fits*

We now consider a clipping of a real low resolution image generated by the Hubble Space Telescope and provided by INAF. This image has been acquired by observing a portion of the sky at high depth, in order to identify a very large number of sources.

Fig. 17 Test 1, rescaling by r_3: Position of the front at time $T = 15.4$ and segmented image, for the FD scheme (27) by using the edge-detector function $\widetilde{g}_2$, with $c_2 = 0.8$

Fig. 18 Test 1, rescaling by r_3: Position of the front at time $T = 14.9$ and segmented image, for the SL scheme (31) by using the edge-detector function $\widetilde{g_2}$, with constant $c_2 = 0.8$

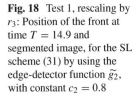

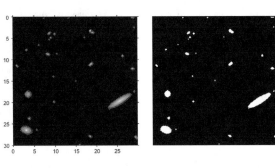

Fig. 19 Test 1, rescaling by r_3: Position of the front at time $T = 15.55$ and segmented image, for the FD scheme (28) by using the edge-detector function g_1, $p = 5000$ and $v = 10^{-4}$

Fig. 20 Test 1, rescaling by r_3: Position of the front at time $T = 28.8$ and segmented image, for the SL scheme (32) by using the edge-detector function g_1, $p = 5000$ and $v = 10^{-4}$

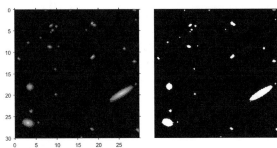

However, this technique leads to an increase in the amount of noise present in the image, as can be seen looking at the left image in Fig. 21.

Let us compare the performances of different schemes with or without a rescaling process. The input images we consider for the segmentation algorithms are reported in Fig. 21. On the left we can see the original image that we store in a file called *real.fits*, in the middle we find the image obtained by using the rescaling function r_1, and the analogous obtained rescaling by r_3 (on the right). Due to the high level of noise, here we increase the time step $\widetilde{\Delta t}$ (from 10^{-4} to 10^{-3}) in the filtering process to obtain the corresponding filtered images. Then we use the rescaling on the filtered images. As can be easily noted looking at Fig. 21, both the proposed rescaling functions improve a lot the visibility of the celestial objects present in the image. The resulting image obtained by r_3 seem to be better.

Fig. 21 Test 2. From left to right: Original image *real.fits*, rescaling of the image *real.fits* by using the function r_1 with $\alpha = 0.25$, rescaling of the image *real.fits* by using the function r_3, with $\beta = 4$

Fig. 22 Test 2.
Segmentation of the image
real.fits provided by the
software *SExtractor*

Let us start commenting the segmentation results obtained by the different schemes without any rescaling process. In Fig. 22, we report the segmentation of the image obtained by applying the software *SExtractor*. In Fig. 23 we can see the performances of the SL scheme (31) for two different choices of the edge-detector function (g_1 and $\tilde{g_2}$) and the SL scheme (32) with edge-detector function g_1. We report only the results for SL schemes since by FD schemes we obtained very similar results. Note that all the different schemes recognize only the two objects visible in the original image reported on the top-left of Fig. 21, so they are far away from the real configuration of celestial bodies.

Therefore we need a rescaling process to improve the results. Looking at the results obtained by r_1, we note that the number of detected objects is highly improved. We report in Fig. 24 the results obtained by the schemes FD and SL only with edge-detector $\tilde{g_2}$, using the edge-function g_1 the front collapses until it disappears from the figure. This is due to the fact that the edges of the objects are very blurred and, even if we choose high values for the parameter p, the variations of gray tones do not allow to detect the presence of an object. On the contrary, using $\tilde{g_2}$ we can find an adequate number of objects, but we cannot detect accurately the boundaries of many galaxies (see Fig. 24). Anyway, this result is more accurate than the performance provided by the software *SExtractor* (Cf. Fig. 22). Looking more in details Fig. 24, some differences between the FD and SL schemes are visible, even if are small (e.g. at the bottom-right part of the big central-upper galaxy we can note a connected part for the FD scheme, which is splitted by SL scheme). For comparison reasons, we report in Fig. 25 the result obtained by the same FD scheme with edge-detector function $\tilde{g_2}$, but the image rescaled by the function r_1 is filtered by the PM method before applying the

Fig. 23 Test 2 without rescaling. From top to bottom: Position of the front and segmented image for the SL scheme (31) by using the edge-detector function g_1, with $p = 10^4$. Same scheme by using edge-detector function $\widetilde{g_2}$, with $c_2 = 0.6$. Position of the front and segmented image for the SL scheme (32) by using the edge-detector function g_1, with $p = 10^4$ and $\nu = 10^{-4}$

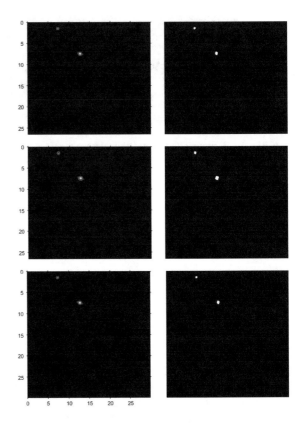

FD segmentation scheme. The position of the final front and the segmented image reported in Fig. 25 show that, applying a nonlinear filtering algorithm as the PM method before the segmentation process, the results can be improved (a lot of small stars are recognized), but a rescaling process is still necessary even if we apply that filtering method.

Finally, let us analyze the results obtained by applying the rescaling function r_3. The parameters chosen in that case is $\beta = 4$ (see the right image in Fig. 21), since greater values provide apparently worst quality. This is due to the poor performance of the Otsu method in this case, note that this method identifies false sources among the pixels of the background. For the type of results provided by the different active contours, similar observations apply to the images obtained by the rescaling r_1, as we can observe from the segmentations shown in Fig. 26. With the g_1 function, the front collapses on itself, disappearing without identifying any object. The results obtained by using $\widetilde{g_2}$ are better even if not satisfactory, due to the noise component, which is excessively amplified, as visible in Fig. 26.

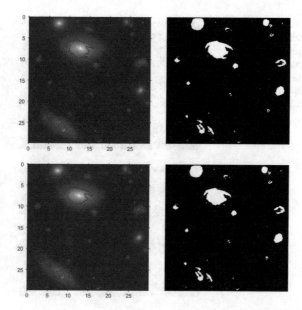

Fig. 24 Test 2, rescaling by r_1. Position of the front and segmented image for the FD scheme (27) (first row) and the SL scheme (31) (second row) by using the edge-detector function $\widetilde{g_2}$, with $c_2 = 0.78$

Fig. 25 Test 2, rescaling by r_1, filtered by 15 iterations of the PM method with f_2, for $\mu = 30$, and $\widetilde{\Delta t} = 10^{-3}$. Position of the front and segmented image for the FD scheme (27) by using the edge-detector function $\widetilde{g_2}$, with $c_2 = 0.78$

5 Conclusions and Future Perspectives

We have proposed different rescaling functions in order to improve the detection of objects in astronomical images, identifying a greater number of celestial bodies. In particular, the use of the function r_3 has improved a lot the visibility, getting better results, in particular for high-resolution images as *f160.fits*. Unfortunately, when the SNR is very low, the results are still not satisfactory, although we notice an improvement with respect to the solutions provided by classical methods without rescaling.

Fig. 26 Test 2, rescaling by
r_3. Position of the front and
segmented image for the FD
scheme (27) (first row) and
the SL scheme (31) (second
row) by using the
edge-detector function $\widetilde{g_2}$,
with $c_2 = 0.6$

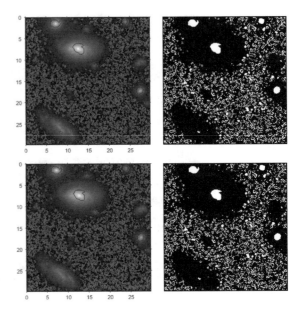

This failure is due to the inaccurate threshold selected via the Otsu algorithm before
applying the rescaling by using the function r_3. For low-resolution images, the function r_1 seems to provide the best results, even compared with those produced by the
software *SExtractor* commonly used by astronomers. Future improvements of the
method can focus on different threshold algorithms to select the optimal one used by
the rescaling function r_3, hopefully this will allow for a correct classification of the
pixels belonging to the background. We also considered a filtering pre-processing
step before segmenting, comparing the linear Gaussian filter and the nonlinear
PM method. What we noted is that the PM nonlinear method improves the results
detecting few more objects, but a rescaling preprocessing is necessary also in this
case, the segmentation fails without it. We have compared the performances of first
and second order FD and SL schemes, using different parameters and two different
edge-detector functions. From the numerical simulations on virtual and real images,
we can conclude that the edge-stopping function g_1 is not a good choice. In fact,
the light sources often have not well defined outlines, so that this function can not
correctly identify them. The edge-detector function $\widetilde{g_2}$ provides the best results, and
is able to detect a greater number of celestial objects. However, these methods are
still not optimal in the case of very disturbed images. In the future, we want to explore
different methods, as for example high-order "filtered" schemes recently proposed
[9–11] or the active contour without edges scheme proposed by Chan and Vese in
[7, 29], able (perhaps) to better identify objects with blurred and not well defined
edges. Moreover, we would like to analyze in more detail the performances of other
filtering methods, in order to find an appropriate choice to reduce the huge amount of

noise that is a typical feature of astronomical images. Some attempts in this direction are shown in [26], they confirm that this will be a difficult task.

Acknowledgements We would like to thank the National Group INdAM-GNCS for the financial support given to this research and the Istituto Nazionale di Astrofisica placed in Rome for the input data. This research has been carried on within the INdAM-INAF project FOE 2015 "OTTICA ADATTIVA".

References

1. Alvarez, L., Lions, P.L., Morel, J.M.: Image selective smoothing and edge detection by non-linear diffusion. II. SIAM J. Num. Anal. **29**(3), 845–866 (1992)
2. Bertin, E., Arnouts, S.: Sextractor: software for source extraction. Astron. Astrophys. Suppl. **117**, 393–404 (1996)
3. Carlini, E., Falcone, M., Ferretti, R.: Convergence of a large time-step scheme for mean curvature motion. Interfaces Free Boundaries **12**, 409–441 (2010)
4. Carlini, E., Falcone, M., Ferretti, R.: Numerical techniques for level set models: an image segmentation perspective. preprint (2018)
5. Carlini, E., Falcone, M., Festa, A.: A brief survey on semi-lagrangian schemes for image processing. In: Breuss, M., Bruckstein, A., Maragos, P. (eds.), Innovations for Shape Analysis: Models and Algorithms, pp. 191–218. Springer, Berlin (2013)
6. Caselles, V., Catté, F., Coll, T., Dibos, F.: A geometric model for active contours in image processing. Num. Math. **66**, 1–31 (1993)
7. Chan, T.F., Vese, L.A.: Active contours without edges. IEEE Trans. Image Process. **10**(2), 266–277 (2001). Feb
8. Falcone, M., Ferretti, R.: Semi-Lagrangian Approximation Schemes for Linear and Hamilton-Jacobi Equations. SIAM (2013)
9. Falcone, M., Paolucci, G., Tozza, S.: A high-order scheme for image segmentation via a modified level-set method (2018). Submitted, arXiv:1812.03026
10. Falcone, M., Paolucci, G., Tozza, S.: Convergence of adaptive filtered schemes for first order evolutive Hamilton-Jacobi equations (2018). Submitted, arXiv:1812.02140
11. Falcone, M., Paolucci, G., Tozza, S.: Adaptive filtered schemes for first order Hamilton-Jacobi equations. In: Radu, F.A., Kumar, K., Berre, I., Nordbotten, J.M., Pop, I.S. (eds.), Numerical Mathematics and Advanced Applications ENUMATH 2017. Lecture Notes in Computational Science and Engineering, vol. 126, pp. 389–398. Springer, Berlin (2019)
12. Fried, D.L.: Statistics of a geometric representation of wavefront distortion. J. Opt. Soc. Am. **55**(11), 1427–1435 (1965). Nov
13. Gao, W., Bertozzi, A.: Level set based multispectral segmentation with corners. SIAM J. Imaging Sci. **4**(2), 597–617 (2011)
14. Hope, D.A., Jefferies, S.M., Hart, M., Nagy, J.G.: High-resolution speckle imaging through strong atmospheric turbulence. Opt. Express **24**(11), 12116–12129 (2016). May
15. Malladi, R., Sethian, J.A., Vemuri, B.C.: Topology-independent shape modeling scheme. In: Proceedings of SPIE Conference on Geometric Methods Computer Vision II, vol. 2031. Springer, Berlin (1993)
16. Malladi, R., Sethian, J.A., Vemuri, B.C.: Shape modeling with front propagation: a level set approach. IEEE Trans. Pattern Anal. Mach. Intell. **17**(2), 158–175 (1995). Feb
17. Micheli, M., Lou, Y., Soatto, S., Bertozzi, A.L.: A linear systems approach to imaging through turbulence. J. Math. Imaging Vis. **48**(1), 185–201 (2014). Jan
18. Mumford, D., Shah, J.: Optimal approximations by piecewise smooth functions and associated variational problems. Commun. Pure Appl. Math. **42**(5), 577–685 (1989)
19. NASA/GSFC Astrophysics Data Facility. A user's guide for the flexible image transport system (fits). Goddard Space Flight Center, 1997. Greenbelt MD 20771, USA, version 4.0. https://fits.gsfc.nasa.gov/users_guide/usersguide.pdf

20. Osher, S., Fedkiw, R.: Level Set Methods and Dynamic Implicit Surfaces. Springer, Berlin (2003)
21. Osher, S., Sethian, J.A.: Fronts propagating with curvature-dependent speed: algorithms based on Hamilton-Jacobi formulations. J. Comput. Phys. **79**(1), 12–49 (1988). November
22. Otsu, N.: A threshold selection method from gray-level histograms. IEEE Trans. Syst., Man, Cybern. **9**(1), 62–66 (1979). Jan
23. Pecci, L.: Metodi level-set per la segmentation di immagini astronomiche. Master's thesis, Dipartimento di Matematica, Sapienza - Università di Roma, Italy (2018)
24. Perona, P., Malik, J.: Scale-space and edge detection using anisotropic diffusion. IEEE Trans. Pattern Anal. Mach. Intell. **12**(7), 629–639 (1990). Jul
25. Roddier, F.: Adaptive Optics in Astronomy. Cambridge University Press, Cambridge (1999)
26. Roscani, V., Tozza, S., Castellano, M., Merlin, E., Ottaviani, D., Falcone, M., Fontana, A.: A comparative analysis of denoising algorithms for extragalactic imaging surveys (2019), submitted
27. Sethian, J.A.: Curvature and the evolution of fronts. Commun. Math. Phys. **101**, 487–499 (1985)
28. Sethian, J.A.: Level Set Methods and Fast Marching Methods: Evolving Interfaces in Computational Geometry, Fluid Mechanics, Computer Vision, and Materials Science, 2nd edn. Cambridge University Press, Cambridge (1999)
29. Vese, L.A., Chan, T.F.: A multiphase level set framework for image segmentation using the mumford and shah model. Int. J. Comput. Vision **50**(3), 271–293 (2002)
30. Welsh, B.M., Gardner, C.S.: Effects of turbulence-induced anisoplanatism on the imaging performance of adaptive-astronomical telescopes using laser guide stars. J. Opt. Soc. Am. A **8**(1), 69–80 (1991). Jan

Printed in the United States
By Bookmasters